THE BUILDER'S GUIDE TO

Running a Successful Construction Company

THE BUILDER'S GUIDE TO

Running a Successful Construction Company

DAVID GERSTEL

The Taunton Press

Cover photo: Metro Image Group

First printing: December 1991
Second printing: December 1992
Printed in the United States of America

A FINE HOMEBUILDING Book

FINE HOMEBUILDING® is a trademark of The Taunton Press,
Inc., registered in the U.S. Patent and Trademark Office.

The Taunton Press, 63 South Main Street, Box 5506,
Newtown, CT 06470-5506

Library of Congress Cataloging-in-Publication Data

Gerstel, David U.,
    The builder's guide to running a successful construction
company / David Gerstel.
        p.   cm.
    "A Fine homebuilding book"—T.p. verso.
    Includes index.
    ISBN 0-942391-36-5
    1. Construction industry—Management. I. Title.
HD9715.A2G46   1991   91-25636
690'.068—dc20   CIP

With love for my parents
Who gave me
Zest for organization and
A willingness to speak my mind.

Also for fellow builders
Fred Blodgett and David Lassman,
Two of the best.

ACKNOWLEDGMENTS

During the four years that I have worked on this book, my wife, Sandra, has patiently listened to and read reams of rough drafts and responded with succinct advice. If I have anything useful to say in this final draft about working with people, it comes in good measure from having watched her skillful and ethical leadership in a business much larger than my own.

Laura Tringali, my editor, has given me the best editorial coaching I have ever received. With refreshing candor she has steadily pointed out my obfuscations and misuse of fancy literary devices, made sure that in rewriting I strove for clarity and vigor, and insisted I keep uppermost in mind the needs of the readers. Have fun writing, she has impressed upon me, because if you don't, your audience will have even less fun reading.

Along with the staff at The Taunton Press, many other people—especially members of The Splinter Group, an association of builders in the San Francisco Bay Area—have contributed to the evolution and expression of my ideas. I have room here only to thank them in alphabetical order, and to apologize to anyone who may inadvertently have been left off the list.

Lou Aggetta
John Altschuler
Harry Attri
Jeff Beneke
Chris Benton
Paul Bertorelli
Marc Brenner
Moses Brown
Curt Burbick
Deborah Cannarella
James Carey
David Chilcott
Bill Childers
Gene Clements
Larry Craighill
Kathleen Curry
Eric Danysh
Linda and Mark Dew-Hiersoux
Phil Emminger
Jeff England

CONTENTS

Introduction   xi

PART 1: A BUILDER'S CAREER

What It Takes   1
Opportunities and Difficulties   7
A Plan   13

PART 2: A BUILDER'S TOOLS

Office   22
Portable Offices   27
Shop   30
Policy Statement   31
Working Capital   35
Insurances: What, Why and How   39
Insurance: Liability   41
Insurance: Worker's Compensation   45

PART 3: KEEPING THE BOOKS

Why and Who   48
From Pile System to Pegboard and Spreadsheet   50
Essentials: Petty Cash and Taxes   58
Adding a Crew   62
Prosperity: Job Costing and Receivables Journal   65
Consolidation and Sophistication   71

PART 4: GETTING THE RIGHT JOBS

Promotion   76
Prospects   81
Qualifying Projects   85
Qualifying Clients   89
Working with Architects   93
Competitive Bidding   99
Price Planning (Negotiated Bidding)   104

PART 5: ESTIMATING AND BIDDING

Setup and Site Inspection	111
Checklists	115
Figuring Costs	125
Marking Up for Overhead and Profit	132

PART 6: CONTRACTS

Why and What Kind	139
The Agreement	144
Conditions	153
Change Orders	160
Subcontracts	164

PART 7: LABOR AND MATERIAL

Hiring and Firing	168
Crew	176
Crew Leaders	182
Pay	185
The Four-Day Week	190
Subcontractors and Suppliers	192

PART 8: PROJECT MANAGEMENT

Safety	200
Job Setup	203
Running Projects	207
Inspectors	216
Wrap-up and Follow-up	219

INTRODUCTION

Since the time I took on my first independent project, I have liked to call myself "builder" instead of "general contractor." I have also preferred to say I run "a company" rather than "a business." "Builder" implies no mere shuffling of contracts and other paperwork, but resonates with pride of craft and stout construction work. "Company" suggests that human loyalties come before legal and financial manipulations.

Now, two decades into my career as a builder, I continue to believe that the company that focuses on the welfare of its employees and clients and on the quality of construction is the company most likely to prosper in the long run. But this is not to say I underestimate the importance of the documentation that goes along with building. On the contrary, I admire the well-crafted estimate, the tight contract, the thorough and accurate change order, the logically sequenced flow of paperwork. I appreciate sound financial management, too. A company needs money to run on, to cover its risks and to meet the needs of its employees and clients. The truest foundations, tightest miter joints and closest attention to

integrity in human relationships will not save a construction company from poor business management.

Unfortunately, there has been precious little available in print, video or seminars to educate the tradesperson-turned-contractor in the myriad skills of contracting. To help fill the void, I have written a guide to the management issues and procedures typically of concern to small-volume builders—those whose companies are small enough that they can give close personal attention to each client and project. New builders just moving over from employment as tradespeople to independent contracting will get a real leg up from studying *The Builder's Guide*. Unlike myself and thousands of other builders, they will not have to cobble together a management program from disparate sources in reaction to their own blunders. Rather, they will have the issues and an approach to dealing with them laid out from the outset. They will have the opportunity to see the need for sound management procedures before costly mistakes make them all too obvious.

If my previous experience holds true, *The Builder's Guide* will also aid seasoned builders.

Gerstel's projects have run the gamut from small jobs, such as this stairway and walk with fence (top), to the recycling of existing structures into new ones (above).

Many have told me they have picked up viable ideas from my articles and group presentations on construction management. I am not surprised, for though I was a successful contractor when I began writing this book, I have learned much from the hundreds of interviews, formal and informal, I have had with other builders as part of my research.

As a small-volume builder myself, I have worked almost entirely in residential construction; my company has built few commercial and no industrial projects. Although we have some experience with the development of "spec" projects, we work mostly with clients who will themselves use our product. For a few years, part of our work load was the rehabilitation of publicly subsidized housing, but the great majority of our clients have been in the private sector (and the same is true for the builders I have interviewed). The contents of this book reflect the dominance of private, residential work over public, commercial and spec construction. A few topics, such as bonds, which are of urgent concern to builders in other niches, are treated only sketchily here.

Readers should also note that *The Builder's Guide* has been shaped by my experience as a builder in the prosperous San Francisco Bay region of Northern California. We have had our construction busts here, but mostly business has boomed during my years in the industry. Moreover, we have seen real-estate values rise at several times the rate of inflation between 1975 and 1990. Property owners have enjoyed huge growth in their equity and have often invested it in large-scale remodeling. Regularly, my company is called for residential remodels costing a couple of hundred thousand dollars and more. Indeed, our "remodeling" projects are sometimes so large that the term hardly fits them. I think of them as "recycling," a type of construction that incorporates an existing structure into a new one. My work, however, also has included many smaller projects, as well as conventional new construction. Therefore, while my experience gives the content of this book a certain spin, I am optimistic that the basic principles will prove widely useful.

For ease of reading, I have organized my discussion of the issues not by abstract categories, but in a narrative form. Like a story, or for that matter, a construction project, my book has a logical beginning, middle and end. New builders will find the material arranged in an order they might sensibly learn it, and will then need to apply it in the operation of their own companies. Experienced builders will find subjects appearing in the familiar sequence—from estimating to project management—they have worked through job after job.

As you go through this book, you will encounter recurring themes I consider crucial to successful operation of a construction company:

• People, not personnel: Good entrepreneurs don't exploit people in an attempt to fatten profits. Take care of the people you work with. They will take care of you.

• Organization: Good builders constantly tighten their procedures for handling every aspect of their work.

• Consolidation: The first tip I got during my carpentry apprenticeship was "Consolidate your motions. Make all your measurements, then make all your cuts." Builders must likewise efficiently group and execute their tasks.

• Minimizing overhead: Builders frequently engage in personal consumption disguised as business investment. They weigh down their operations with unnecessary real estate, staff and equipment. The high cost of such ego toys can crush a company, especially when work slows. Builders must avoid nonessential overhead. Travel light. Run lean. Keep overhead down.

• Safety: Builders who promote safety are practicing enlightened self-interest. They run a good company that is also a sound business. Good safety policies are among the most vivid examples of why concern for people is financially productive.

Other builders might stress different themes. Every reader will invent improvements on the ideas I offer here. But it is also true that my way works. The practices I describe in *The Builder's Guide* consistently support the construction of sound and attractive buildings. They leave clients so satisfied that they drum up business for our company among their acquaintances. They create enjoyable work environments and good incomes for the company's employees and subcontractors. To myself, they have brought satisfaction, opportunity and prosperity beyond my imaginings when I first picked up a hammer.

Among the advantages of running a well-organized company—you don't rush about madly all day trying to keep the thing patched together. Here Gerstel settles down for his customary long morning coffee break to peruse the latest returns from the National Basketball Association.

Note:

The Builder's Guide covers several areas of concern to builders, including accounting, taxes, insurance and contracts, that are subject to federal and state laws. In each case, the material presented in *The Builder's Guide* is intended only as an introduction to the subject. It cannot substitute for legal counsel. In all matters that involve local, state or federal laws and regulations, builders should seek additional guidance from government agencies and an attorney.

A BUILDER'S CAREER

PART 1

What It Takes
Opportunities
  and Difficulties
A Plan

 ## WHAT IT TAKES

Builders I know include among their previous occupations homemaker, shipyard executive, art professor and pinball-machine vendor. Some come to construction for romantic reasons. One contractor, even after 30 years of building, will stroke a piece of kiln-dried molding and croon to a client, "Look at that grain, tight, like the leaves of a Bible." Some come with dollar signs in their eyes. But regardless of the initial motivation, to succeed, all have had to develop a considerable range of skills and aptitudes.

At the foundation is a sound basic education. Builders must read with ease, write clearly and grammatically and execute basic business math quickly and accurately. Some people graduate from high school with the needed education. Some have it after two years of college. And many still do not have it after getting a BA degree, or even an advanced degree. But have it you must.

As a builder, you read complicated legal and technical texts. You write proposals, revise contracts and compose change orders. For estimates, you organize pricing structures and formulas, and calculate quantities and costs. To understand your company's financial performance overall, as well as job by job, you must analyze spreadsheets. If you cannot readily learn and handle those tasks, you will tend to avoid them. "Too much of a hassle," you will say.

For a time, you may slide by even while skimping on your reading, writing and arithmetic. But eventually you will get clobbered. For example, a builder for whose case I served as an expert witness had operated 10 years without creating good estimating forms or

Skills and Aptitudes Needed by Aspiring Builders

Before starting out in their own businesses, aspiring builders should ask if they have what it takes to run a successful company:

- Sound education in the "three R's"

- Trade skills

- Commitment to developing management skills

- Organizational rigor

- Flexibility enough to roll with the punches

- Passion for building

contract documents. Then he began a large remodeling project on the basis of a handshake with the owner. As the project expanded, he wrote no change orders and was lax in billing. With the project two-thirds complete, he handed the client a bill for $20,000, bringing his total charge to $60,000. The client refused to pay, alleging the charge was far in excess of the numbers mentioned in the contractor's original verbal estimate. Without paperwork to show otherwise, the builder had no choice but to sue. He spent two years in litigation pursuing the 20 grand, and succeeded in winning only a third of it. If he'd had proper documentation and management procedures in place all along, litigation would not even have been necessary.

Beyond the "three R's," builders need a thorough knowledge of the trade. You find exceptions, but most successful general contractors have achieved journey-level competence or more at carpentry. They have smashed their thumbs, bruised their skulls and trudged through mud. They have pumped concrete, framed walls and coped crown molding. Because they have been in the trenches themselves, they enjoy the respect of their carpenters and can effectively supervise them. Because they have done their time with the tools, they can recognize the moves of a good carpenter—and an incompetent one.

Builders who do not have firsthand trade knowledge feel the disadvantages. One I know attempted to become a general contractor after success as a painter. He promptly lost $30,000 on a remodeling project and went bankrupt. Looking back on his failure, he remembered that he had lacked confidence in his estimates. Although he made them with a powerful computer program, the numbers generated felt abstract because he could not verify them against personal experience. When it came to the construction, he found that the logistics were completely overwhelming. "You must," he said, "have a knowledge of the trade so that you can worry about management without struggling with building itself." A builder who was largely self-taught as a carpenter has this to say about the lack of trade experience: "It's a drawback in several ways. There are standard ways of doing things just as there are standard dimensions in wood. They have become standard because they are the most efficient way to do things. Because I don't know those approaches, I have to discover them as if I were the first person coming across the problem. I'm always reinventing the wheel. That has affected my confidence level. I feel uneasy telling clients I can do things."

But aspiring builders need not only carpentry skills, they need to learn them from professionals. The best setting is a well-run small-volume company. As an employee of a big firm that handles large projects, you are likely to be assigned the same tasks repeatedly. I have known carpenters who had built tract or apartment housing for

A Letter Written by Author to a Young Man Hoping to Become a Builder

After reading an article by Gerstel in *Fine Homebuilding* magazine, a young man with a couple of years of trade experience and another year at a technical school under his belt wrote to ask for advice on setting up a construction company. Gerstel wrote back to discourage him from starting too soon.

Steve Smith
862 High Mountain Road
Brewster, NY 10509

April 10, 1988

Dear Steve,
Thank you for your nice remarks about my article. You ask if I have any suggestions for you. Yes I do. DON'T START YOUR OWN BUSINESS YET. Master your trade first. Do whatever it takes. Work for lower wages. Commute great distances. Move to another state. But get yourself a job in a well-organized small company where you will learn sound construction techniques and be exposed to good project management.

Work for that company (or companies) for three to four years. In your last year of employment, set up all the systems you need to manage a construction company. Once you have them in place, not before, begin seeking your first projects….

The reports vary a bit, but they all say that at least nine out of ten contractors fail in business within the first couple of years. Two reasons they fail are that the people who start them don't know their trade well, or they don't know business management. If you start your own company now you will know neither. It takes more than two years in the field to learn a trade. And it takes five years to become a good manager of your own company. You do not want to be still unsure of the trade when you take on the challenge of management.

Steve, you will get where you want to be as a builder sooner if you don't rush.

Good luck,
David Gerstel

years, but learned only a single skill, such as rolling joists or laying subfloor. In the employ of a larger company, you may be shipped from project to project as each in turn requires your specialty and rarely see construction from start to finish. By contrast, in a small-volume company like my own, each carpenter participates in all tasks, from demolition to trimming out the completed structure. With the smaller company, you will likely work through the whole construction sequence on one project after another. Over and over you will see the meshing of all the subtrades with rough and finish carpentry. You will learn how a building goes together. That's invaluable, because when you become a contractor, your responsibility will be to see that the whole building *does* go together.

Ironically, after working years to learn carpentry, you must put aside your tools if you are to organize a construction company. True, you can operate independently without giving up hands-on work. You can bid and contract for only small-scale projects, then build them yourself. Much pleasure can be had from such a compact enterprise. But you are working as an independent artisan, not operating a company.

To execute larger projects crisply, a builder needs to put the tools aside, organize a company and make management the priority. Here is the compelling logic: Effective handling of a project, such as a new home or large addition, requires a stable crew who have learned to work as a team. For a crew to be stable, it must have steady work. Providing that work—developing customers, estimating and negotiating contracts—while tending to all the other necessary office and field management is a full-time job. Builders who also attempt to lead the on-site carpentry are forced to neglect crucial management responsibilities.

In my own career, I recognized too slowly the importance of management. For years I worked with ad-hoc crews and pushed management tasks into the leftover, late, tired hours. And everything suffered. My trim joints met at an offset, for as I made them my thoughts drifted to the estimate for my next job. My employees loafed and sulked, but I had to put up with them, because I did not have the time to hire anew and hire right. My paperwork lapsed. I was too busy building to track expenses, so I never knew what a project was costing until it was over and I had added up the receipts. I made my bids fit every project that came along, for I did not have time to pick and choose; I earned far less than I could have with more selective bidding. Always I was tense and tired.

Finally, as my projects grew bigger, I saw that I was relegating myself to apprentice work at the job site: hauling lumber, cutting blocks, sweeping. I could not handle more complex carpentry tasks. Too many people called for my attention. Too many management responsibilities pressed at my mind. Finally, I decided to break the habit of strapping on my bags at the start of each work day, promoted my best carpenter to crew lead and began a progressive withdrawal from the job site to concentrate on running my company.

Builders resist committing to management for a variety of reasons. After earning their keep with their tools for years, they imagine they will go broke if they put them down. They don't notice that all around other builders prosper, having long ago traded their tool bags for a briefcase. Alternatively, they undervalue management. They are proud to be wearing those tools. They even brag to clients, "I do the work myself" (unlike those slick characters who take their customers' money for just shuffling paper). They don't see that on any but very small projects, management is the work most crucial to success. Every other element of the project depends upon it.

Finally, builders resist managing because they love carpentry and find management a chore. As one says, "The fun part of being a contractor is the hands-on work." The man who first employed me as an apprentice feels that way and has chosen not to run a company. Instead, he has become the consummate woodsmith, operating out of a spacious shop with his home above. Much of the time he works alone. Occasionally he hires a helper or partners up for a larger project. He seeks only the most challenging work, earns good money and is happy. Such a position of independent artisan is the sort you might strive toward if hands-on work is for you the only satisfying part of construction. But even then you must tend to certain essential management tasks. Neglect them and you lose money,

Chris Benton, owner of White Hound Construction, once employed ten carpenters. But he has been much happier since he returned to a one-person shop, doing the custom jobs he first fell in love with as a teenager.

peace of mind and client loyalty. The systems we will discuss in coming chapters will help you compress those tasks into a few effective hours each week so that you need spend little time away from your main love.

Management, too, can be satisfying once you become proficient at it. Builders so frequently refer to management tasks as "headaches" and "hassles" because they lack the techniques to perform them well. But the trades are also frustrating if you don't know what you're doing. Once you have mastered management, the work will seem as fundamental to good construction as building plumb, level and square. When you pass a job site, signs of poor management—debris spread over the lot, fitful progress, idle and glum workers, poor sequencing of trades—will hit you the way sloppy joinery jolts your carpenter's sensibility. You will take pride in your skill at running organized projects. You will experience a thrill of leadership that is absent from the independent artisan's operation.

Once you have committed to running a company, you will be challenged to develop organizational skills far beyond those you required as a tradesperson. Carpenters must work in an organized way, consolidating tasks and executing them in efficient sequences. But their tasks span a few hours or days. Builders organize for the weeks ahead, and the months and even the years. They constantly plan, making maps—charts, schedules and lists—of where they are going. They do not let matters fall where they may. In *Professional Remodeling Management* (see Resources, p. 223), author Walt Stoeppelwerth suggests that in the poorly run company the typical answer to every question is "We'll play it by ear," "We'll work it out," "Piece of cake" or "No problem." When I hear those flippant replies, I grimace in anticipation of a marginal performance. The competent contractor will say instead "Let me make a note of that," or "Let me put that into the schedule."

Paradoxically, the imposition of order on the flow of work by a builder must be accompanied by another and opposite trait: flexibility. As you move from on-site work to running a company, you must delegate to others the tasks you once performed yourself. Effective delegation requires the flexibility to let others do the tasks in their own way, not exactly as you would prescribe. You destroy people's motivation and rhythm by forcing them to move in your patterns rather than their own.

As a builder, you must be able to adjust not only to people but to events. You must be able to roll with the punches, a lot of punches. Good builders organize in the knowledge that their plans will be regularly blown apart by weather, illness, change orders, economic recession, turnover and hot opportunities. When disruption comes, they flow with it. One builder I admire lost his estimator, office manager and top three leads in a span of months. I expressed my concern for the health of his company. "Oh, the changes have moved us forward," he responded brightly. "Your best people build your company, but they also define and limit it. When they leave, you have the opportunity to stride off in new directions with new people." Of course, as a well-organized builder he had a deep file of prospects he could call to replace his departing employees.

Builders are more than managers; they are entrepreneurs who create and win projects. They must be able to sense opportunity in the marketplace, as successful entrepreneurs in general understand. In one study, a group of businesspeople of diverse experience was asked to select from four factors the one both necessary and sufficient for starting a company. Most responded with "money," "hard work" or "an idea or product." But the entrepreneurs chose "customers." They understood that if there is no market, no amount of capital investment or effort will move even the best-made goods.

One final attribute a builder must have is passion. Among entrepreneurs, it is axiomatic that to do well in an enterprise you must love it. In building, if you do not, your indifference will result in shoddy management and shoddy construction, spawning that whole litany of disasters from failed inspections to upward-spiraling insurance premiums, labor-board investigations and lawsuits that are so depressingly common in our industry.

Passion. Entrepreneurial talent. Flexibility. Organizational rigor. Commitment to management. Trade knowledge. Sound education. Do you need it all to make it as a builder? Not to begin with. But what you lack, you must cultivate. The construction industry is competitive; the failure rate, as we shall see, is huge. But for those men and women who are willing to acquire what it takes, the industry offers ample compensation.

OPPORTUNITIES AND DIFFICULTIES

The U.S. economy generates great demand for builders. A growing population, changing technology and the toll of time require continuous new construction and rebuilding of existing homes and commercial structures. As of the late 1980s, the nation was investing approximately $120 billion annually for construction, with two-thirds going to remodeling or renovation. With its hidden conditions, variety and resistance to routine production methods, remodeling especially favors small-volume builders. You can give each project close personal attention, adjusting and innovating as work proceeds, thereby creating profit for your company. One remodeler's consultant with a national clientele claims that a well-run remodeling company will find itself in demand anywhere in the country. Small-volume builders find a niche as well in the various other sectors of both private and public new construction.

As a small-volume builder, you can earn a high income, though it will not come right away. For the chance to establish a track record in your community, you may at first have to work at below-market rates. And, as you start up, you will be slow at management tasks. As a result, you may well work longer hours for the same income (or even for less) than you earned as an employee. But as your reputation grows and your managerial and entrepreneurial skills sharpen, your income can rise to the level of successful professionals in other fields. As I write, in the spring of 1990, a small-volume general contractor can earn a high five-figure and even six-figure income annually, or far more by moving into development.

Lean years do punctuate the lucrative ones, but nonmonetary rewards arrive in both. As one contractor says, "The buck stops with you, and sometimes it stops in your pocket," but running your own company offers more than money. It is a vehicle by which to express your values to your community. If you cherish good craft, for example, you can make decisions that foster enduring and architecturally vigorous structures. "This is the way to build," you proclaim publicly. If the vision of a sustainable economy galvanizes you, you can turn your job sites into virtual recycling centers, like the builder I know who takes pride in running a job so that nothing but a little scrap drywall and a handful of sawdust go to the dump.

Among the best rewards is the appreciation of clients for work done well. Sometimes it comes indirectly. Your phone rings and the caller says, "Our friends the McFeely's say you did a great job for them and made us promise to talk with you before we selected our contractor." Among my own favorite moments in construction was an accidental encounter with a client in a coffee shop. A year earlier my company had added a master bedroom and bath to her small home, giving her a sanctuary from two teenagers she was raising alone. "I want you to know I'm happy in my house now," she said. "I like being home. I hated being there before."

As your career lengthens and your projects dot the town, they anchor you to your community. You have built its shelter. "That's why I like building," says Ann Hollingsworth, owner of Rosewood Construction. "I like making something. I like imagining it in my mind and then figuring it out and doing it and then it's there. Ten years later you come back and it's still there doing its job." But while opportunity for rewards are abundant, they do not come easily for the small-volume builder. In fact, few builders survive long enough in business to enjoy them.

Contractors move in and out of business at a startlingly brisk rate. Robert Swatt, a San Francisco architect, finds that working with builders can be a little like going out to eat. He tries a restaurant, enjoys a meal, then goes back and discovers the restaurant's windows papered over. Likewise, he is impressed by a builder during one project, calls him or her to request a bid for another and gets a "phone disconnected" message. On the average, say various surveys, of 100 builders who hang out a shingle, only five will be in business after a couple of years.

By no means do all those start-up builders "fail" in any brutal sense of the word. Many simply quit because they don't like contracting and opt to work again with their tools for a wage. But too often their young companies do actually collapse. Construction businesses are among those most prone to lose control on the steep curves of the business cycle. When markets boom, builders often overexpand, then find themselves lacking either personnel, capital or the management systems to handle the enlarged work load and are crushed by cost overruns. On the other hand, when business sags because of high interest on construction loans or nationwide or regional recession, work slows to a trickle. With no money coming in, builders cannot meet the obligations they've taken on in the heady years of rising markets. I have known small-volume builders who made well into six figures one year. During the next, they were unable to pay the installments on their trucks or the mortgages on their homes.

Builders who have come up during a boom period, such as the one enjoyed in many parts of the U.S. during the 1980s, often don't see the slump ahead. As they luxuriate in their companies' rapid growth, they ridicule warnings that the go-go years are always numbered. But seasoned builders know to operate during the good years with an eye toward bridging the inevitable lean times (Travel light! Control overhead!). No management procedure, however, can take all the risk out of building. You never know, for example, when recession will hit—it can come just after you have taken on the new overhead required by seemingly prudent expansion. And while the business cycle is no more controllable than the weather, even your own operation and projects cannot be completely tamed. Examples of unforeseeable trouble:

- In the middle of his peak season, a builder experiences a spinal deterioration. While waiting for and recovering from surgery, he must read plans, write change orders and otherwise supervise his projects while flat on his back.
- Half of a builder's leads quit in the same week. While he searches for replacements, costs race upward on his projects.
- During final cleanup on a kitchen remodel, an apprentice uses a nylon pad to scour the cabinetmaker's crayon numbers off the plastic-laminate doors and panels. His conscientious scrubbing dulls the gloss, requiring that the doors and panels be replaced at a cost exceeding the entire markup for overhead and profit on the project.

Along with the economic risks of building come psychological burdens, such as feeling low on society's totem pole. I am among many builders who do not share that feeling. I fancy myself a member of one of the most independent of professions and part of the ancient and distinguished tradition of providing shelter for one's community. But I often hear good builders complain of their lack of status even as compared to other construction professionals. "It really fries me," says one builder with a national reputation, "that engineers and architects are seen as professionals, and we are seen as sleaze."

Compounding the discomfort is the unfortunate fact that the public's negative perception of contractors springs from certain hard truths. Woefully low standards often do prevail in our industry. In my own area, drywallers call their basic product—textured ½-in. wallboard—"rape and tape, blow and go." For other trades, the usual installation evokes similar descriptions. While standards are low for the actual construction, for entrance into general contracting they hardly exist. The gates to our profession, if indeed there ever were any, have long ago been knocked right off their posts to be left rusting in the gutter. The National Association of Home Builders reports that typical new contractors are unemployed construction workers trying to create jobs for themselves (such was the case for myself and many other builders I know). We take state examinations that require only superficial knowledge of code requirements and safety and have little or nothing on estimating and business practice. We then grope our way into the marketplace, typically to do damage to ourselves and our clients. Of course, by dint of hard work, many of us lever ourselves upward and become competent managers, entrepreneurs and builders. Nevertheless, we continue to be tarred with the same brush used on the mediocre and incompetent members of the profession. "You just don't build like they did in the old days," we are regularly informed by sidewalk superintendents speaking with a certainty that can only be born of vast ignorance.

To add to the pressure builders feel, though the public denigrates our trade, it also unwillingly creates powerful incentives for shoddy work. While on one hand complaining of poor construction, on the other, as *Forbes* magazine (July 14, 1986) points out, Americans crave "bigger, more lavishly equipped homes than any other people on earth, or in history." The demands for size cannot be met without massive corner-cutting. The smart builder will urge customers, "Build less and build better. The well-constructed smaller home will give more satisfaction and be a sounder investment than a big piece of junk." Some customers welcome the guidance. But many are afraid of "too good a job," especially if they plan to sell soon. "Americans want the sizzle, not the steak," an architect once told me. But he had it only partly right. Often they want a sizzling steak for the price of just the sizzle. When the sizzle dies out, when their cheap buildings begin to pull apart at their thousands of seams, they complain loudly. Often, these days, they sue.

The Perils of Apprenticeship

Though he had no experience in construction, Tom had always admired good architecture, and when his other career plans didn't work out, he decided to become a general contractor. Realizing he first had to learn to build, he quickly found a job helping with the construction of a greenhouse. Soon after starting, Tom learned that at the clients' request, the contractor was building the greenhouse half on the public parkland adjacent to the clients' lot. The clients figured that by the time the park managers noticed, their greenhouse would have been in place so long they would be able to claim the parkland by right of eminent domain. The contractor did not mind. His main concern was making sure he had a few lines of cocaine ready for dessert after lunch each day.

When Tom had been inspired by articles in *Fine Homebuilding* magazine to pursue his new career, he had envisioned training under a different kind of boss. He gave the greenhouse builder notice and moved on. But at his next job, the builder paired him with another neophyte carpenter, gave them a load of lumber and a few sketches of a deck, and told them to build it. If they needed advice, they could reach him aboard his sailboat down at the bay.

Tom moved on again. This time he got a job with a developer who was converting a warehouse into shops and offices. When the developer discovered Tom could read, he handed him the plans and told him to take charge of the carpentry. "It was a circus," Tom later recalled, a dangerous circus with untrained adolescent laborers running wildly through the site trying to out-macho one another.

Tom moved on, and with his next job he thought he had finally struck gold. He got on with an architect/builder putting up wood-sided homes in a sylvan hillside neighborhood. But on Tom's first day, the builder dropped at 2x12 on his own head. When he returned from the hospital he called the crew together. While getting his stitches, he realized that the project was going too slowly, so he would now offer the crew bonuses for increased productivity. Shortly afterward, an apprentice chasing the bonus tried to rip a 2x4 with his Skilsaw while holding the board straight out in the air. He cut off a finger. So it went, until the project finally reached completion a year later, eleven months behind schedule.

Tom next found employment with another architect/builder, this one doing a remodel and addition. The project soon soared over budget, the client fired the architect and Tom was laid off. Then Tom hit bottom, signing on with a remodeler named Prokovich.

Shortly after Tom went to work for Prokovich, they were faced with installing a fiberglass tub/shower unit for the bathroom that would not fit through the doors of the house. Tom suggested sending it back for a knock-down unit, but Prokovich said no. Two days later, the crew had the unit in the bathroom—having built a ramp to the second story, demolished an upstairs window, broken out a stucco wall and cut through plaster, wiring and water-supply lines—only to find that the unit was the wrong size. As Prokovich took out a saber saw to cut it down, Tom decided to seek another profession.

The high rate of discouragement and failure among builders occurs in part because they must often organize their companies from a wobbly base. In our demoralized educational system, mere perseverance does not guarantee a good grounding in the "three R's." I once knew a superb lead carpenter who aspired to be a contractor, but despite three years at a state university he could not even correctly spell the names of the trades he supervised.

Once you enter the trades, because standards are so low, you often don't get adequate exposure to sound building practices. Even with a relatively capable company, an apprenticeship can be grueling. The work is hard. You bruise. You bleed. Often you feel humiliated. Says one contractor who started his own company earlier than he should have, "I got tired of being treated like a Skilsaw, like a disposable resource. My only way out was to start up my own outfit."

Women trying to break into construction face the additional obstacles of discrimination when seeking employment in the trades, and harassment when they find it. Sexism continues to dog women when they make the move to contracting. Builder Mimi Ward relates that at meetings she attends with her business partner, who also happens to be her husband, she is always introduced as his "lovely wife." All the women, she says, "are introduced as lovely wives. We don't have names."

The greatest impediment to establishing a solid foundation for the company you aspire to start is the unavailability of management training. With unusually good luck, after you have learned your trade, you may land a staff position in a stable firm. You can then observe and learn firsthand the procedures for marketing, estimating and bidding, contract writing, accounting and project management. But such opportunities are rare. For the most part, builders are by necessity self-taught managers. They have had to scrape together information from other struggling contractors, from scattered courses and seminars and from sketchy articles and books. Often they are prodded forward only by mishaps and fear. You hear of another contractor receiving a huge bill after an IRS audit, so you hurriedly set up a legitimate payroll and organize your books. You underbid a job so badly you work months at minimum wage to build it, so for the next project you set up an estimating checklist and begin tracking job costs. A client refuses to pay you your last $16,000, so you assemble a thorough contract and change-order procedure.

Such fumbling, panic-driven development is inefficient, and it is dangerous. Looking back on the earlier years of my own career, I can see myself crossing a lake of ice so thin it broke behind me even as I took each additional stumbling step forward. Only from greatest luck did that ice not break directly under me before I reached shore. I can see that with a different approach, I could have progressed more firmly and rapidly, given better service and enjoyed my work more. I could have planned my business, learning about and putting in place the systems it would need before starting up.

 A PLAN

Everyone has a wish list. Mine includes ending racism and sexism, mastering my left-hand hook shot and persuading one out of ten readers of this book to create a business plan. The last wish likely enjoys the least chance of being granted; builders are just too accustomed to the seat-of-the-pants approach. "You can't wait until you've got all your ducks perfectly in a row," scoffs one. "You've got to respond to opportunity. You may have nothing more than your truck and a few tools. But when opportunity comes, you pull together the resources you need and do the job."

Exactly that attitude, I propose, accounts for the high failure rate in contracting. Builders do need the flexibility to respond to the opportunity of the moment. But good planning and organization increase flexibility, because they put in place the systems needed to handle opportunity. If you do not have the systems, that shimmering oasis you chased down the road can turn out to be a bog that sucks your company under. Start-up builders should plan the initial organization of their companies before taking on their first customers. Established builders should plan anew before each expansion or change in direction. And builders whose companies have grown haphazardly should systematically examine and retrofit each aspect of their operations.

Builders do not need a "business plan" in the usual sense of the term—a slick presentation of market studies and volume projections such as other kinds of enterprises put together to woo bankers or investors. Builders do not need investors. You can save the money to start a small-volume construction company out of your wages. If you run your company properly, it will generate sufficient new working capital to fund its own expansion. Unless you want to do "spec building," which requires short-term construction loans, you will not need to borrow.

What you do need is a step-by-step plan such as that shown on p. 14 for setting up your company. A plan that fits your needs may give you less or more time to set up your company. It may put tasks in a different sequence. What matters is only that you do have a plan with an orderly and complete sequence, and that you act on it. Nevertheless, any good start-up plan will likely include most of the steps in the sample plan. In the remainder of this chapter, I will briefly describe those steps, saving detailed discussions for later chapters. For example, here I will only mention the need to acquire proper worker's-compensation and liability insurance. On pp. 39-47 I will discuss in detail the crucial aspects of insurance with which a small-volume builder must grapple.

A Sample Business Plan

Write your plan on a single large sheet of paper so that you can see it all at a glance; leave room to write in additional steps. Create a plan to suit your particular situation using this one as a springboard.

Preliminary:
Self-evaluation of the six skills and aptitudes needed by builders. Bring all to acceptable start-up level.

Twelve-month organization of basic business system:

First month:
• Obtain Federal Identification Number from IRS
• Open a business checking account
• Set up a working capital account and establish a goal
• Select an accountant and attorney
• Select form of business (sole proprietorship, partnership or corporation)

Second month:
• Select niche (first and second choice)
• Join a builders' association
• Research market

Third month:
• Organize a home office
• Organize a home shop and storage

Fourth month:
• Write a company policy statement
• Sign up for the state contractor's-license exam

Fifth and sixth months:
• Set up bookkeeping and payroll systems
  a. Research systems and tax-reporting requirements
  b. Purchase the bookkeeping systems
  c. Set up systems with a review by an accountant

Seventh and eighth months:
• Create an estimating system
  a. Research
  b. Set up
  c. Practice estimating

Ninth month:
• Develop legal documents
  a. Contract agreement and conditions
  b. Change orders
  c. Subcontract

Tenth month:
• Insurance
  a. Research worker's-compensation and liability insurance
  b. Select an agent to provide future policies

Eleventh month:
• Study building codes, lien laws and other pertinent contracting law
• Take the contracting exam
• Set up a portable office

Twelfth month:
• Develop a phone file
• Obtain a local business license
• Order business cards

Second year:
• Update business plan
• Promote business
• Begin contracting for projects
  a. Side jobs
  b. Independent artisan
  c. Hire crew
  d. Focus on management, refining systems and expanding company

As a preliminary step in organizing a company, aspiring builders should take inventory of their skills and aptitudes. Ask yourself if you have what it takes (see p. 2). Can you read easily and with comprehension? Can you write clearly and grammatically? Are you skilled at basic business and builder's math? Have you learned your trade—not in school, not in theory, but hands-on in the field from professional builders? Do you have a passion for building, a commitment to management, flexibility and entrepreneurial flair? If you lack any of the "three R's" or trade skills, get them in school or in the field, respectively, before you start your company. If you lack passion for building or commitment to management, give this book to a friend and set your sights on another career. If you lack flexibility and entrepreneurial moxie, count on acquiring them, most likely with some aches and pains.

If you do have what it takes, you can begin setting up a company. In the first month, open a business checking account to separate your company's income and expenses from your personal finances. Your business account will also provide the data you need to analyze your company's financial performance and to complete tax returns. Set up your business checking account at a bank that is designated a federal depository—that can, in other words, accept the payroll-tax deposits you will be making when you hire employees.

Along with the business checking account, you may wish to establish a working-capital account. New businesses frequently fail for lack of adequate capital; your company must have money in the bank. As you plan your operation, you must project your needs for working capital, as well as the monthly amount you must set aside to be adequately capitalized when you contract for your first projects. For simplicity's sake, you may prefer to keep working capital in the business checking account. But by opening a separate account, such as a money-market account at the same bank, you will be able to draw higher interest. As needed, you can transfer funds from your capital to your business checking account.

Also in the first month, find an accountant and lawyer. Get references from builders you respect for their management skills and ethics. Interview several accountants and lawyers, and choose people you like and who answer your questions clearly. An accountant or lawyer who insists on using incomprehensible fancy language is of no use to you.

You may wish to get your accountant's and lawyer's advice before taking the next step, determining the form of business organization you prefer for legal and tax purposes. You have three choices: a sole proprietorship, a partnership or a corporation.

Most builders start out in sole proprietorships, the great attraction of which is simplicity. You automatically become a sole proprietor when you contract for your first independent project, be it a house or a fence. You are now in business. And you are the sole owner, or proprietor, of the business. If you elect to operate as a sole proprietor, legally speaking and for tax purposes, you *are* the business. Its earnings are considered your earnings by the IRS: You pay personal income tax on the entire earnings of the business, not only on the portion you take for personal use. Likewise, you and your business are one and the same before the law. You can be held personally liable for its obligations. If for any reason you are sued and lose, the plaintiffs may be able to lay claim to your personal assets—your savings, investments, even your home.

The Wheel of Tasks

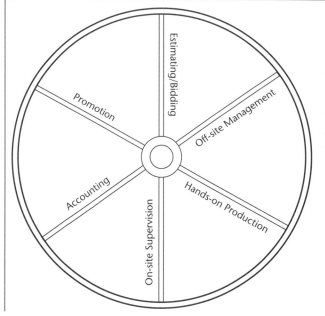

Traditionally, the management of a company is diagramed as a pyramid or other hierarchical form. In the small-volume construction company, many of the tasks are likely to be performed by one person — the owner. Then the tasks are mutually supportive, like the spokes of a wheel. Accounting contributes to estimating. Estimating sharpens on-site supervision. On-site supervision offers opportunities for promotion, and so on.

The second business form, partnership, often appeals to new contractors. It is easy to feel intimidated by the complex cycle of tasks that must be performed in a small-volume company, and partnership may appear to offer shelter from the storm. Unfortunately, partnerships often add to, rather than protect against, the troubles of starting a business. Rather than supporting one another, partners often conflict. Both are set back. Embark upon partnership only after careful consideration. Avoid the classic pitfalls:

- Don't take a partner because you are scared (however reasonably) of running your own company and want to be holding someone's hand when you take the big step.
- Don't partner up to compensate for ignorance. For example, don't take a partner (including a spouse) to do the office work because you are unable to handle the books. You must be competent in all areas of your company's operation, because to be truly competent in any area of office or field management, you must understand how it works in relation to all the others. To manage a project, you must understand the bidding that precedes it and the job costing and bookkeeping that accompany it.
- "Don't," to quote an old entrepreneur's axiom, "take on a partner to do what you could hire done."

Form a partnership only to bring into your company someone with skills you must have who will not settle for employeehood. Partnerships succeed when the partners share a common sense of purpose and ethics and feel their skills are complementary. While each partner could manage all the company's tasks, they also understand their respective strengths and share or divide functions accordingly. The one who works in the field is not resentful of the one who stays in the office on rainy days and schmoozes with the customers. When the sun is shining, the office partner does not grow jealous of the partner who is outside getting a tan. If you are suited to partnership, the right partner may prove a great advantage. One successful partner I know told me you can get more than twice as much done as when working alone.

Ordinarily, partners' tax and legal positions parallel those of sole proprietors. They split their company's earnings, and each pays personal income tax on her or his entire share. Both can be held personally responsible for the company's obligations.

Often, to improve either their tax or legal positions, sole proprietors and partners incorporate. After incorporation, the builder (or builders, in the case of a partnership) and the company are taxed as separate entities. The builder draws a salary from the company and pays personal income tax on it. Additional earnings stay in the company accounts and are taxed as corporate profit. Sometimes (depending on the state of the tax code, which has fluctuated wildly in recent years) a builder who incorporates can enjoy a variety of at least momentary tax advantages. For example, the combined tax on your salary and your corporation's profits may be less than the tax you would have paid as a sole proprietor. On the other hand, if you do eventually draw the profit out, perhaps as a bonus to yourself, you may have to pay personal income taxes on the same money that was already taxed as profit.

Because the tax advantages are not great, at the present time small-volume builders incorporate primarily to separate themselves legally from their companies. After incorporation, the company is considered a separate "person," having distinct legal obligations and liabilities. If a client sues, the company's assets may be vulnerable, but your personal investments and property are not necessarily up for grabs. Exactly how much protection the corporate veil provides for personal assets is, however, a matter of debate. Some lawyers insist that proper maintenance of a corporation virtually guarantees protection of personal assets. Others will tell you the veil can be readily pierced. At the least, however, as one veteran contractor says, incorporation does seem to "set up an additional barrier between your life savings and litigious clients, who think they should take everything you have because you made a real or fancied mistake on their project."

The protection of incorporation naturally has a price. Incorporating costs money. And once incorporated, you must keep records, hold meetings and observe other legal formalities that may make you nostalgic for the simplicity of sole proprietorship. If you neglect the formalities and maintain a "paper corporation," it will shrivel in the first heat of a lawsuit.

Whatever business form you choose, as you plan your business, you need to determine the niche you intend to occupy. This is one of the tasks to be accomplished in the second month of your 12-month organization. The size of your original niche will depend on your trade skills. Though it may grow as the company matures and your skills expand, you will find competing in more than a few areas of the construction industry difficult. I define my niche as "intimate residential construction"—working closely with clients who want to renovate, remodel or build a home. When we venture outside that niche, even into related fields, we have a tough time holding our own against the specialists.

After determining your niche, join an association of builders—ask around for the best in your area—that meets regularly to discuss construction issues. Start-up contractors do not typically join such groups. But membership can speed you toward understanding and dealing with the management issues of concern to a construction company. Moreover, the veteran builders you meet there will be a prime source for the market research you should also do in the second month of organization. Before entering into business, determine whether there is room in your intended niche. You may want to specialize in kitchen remodeling. But if the builders at your group tell you there are already too many outfits in town reaching for that slice of the construction pie, you can consider other directions.

In the third month, having aimed toward a feasible niche, you are ready to create a base of operations—a shop and an office. We all love a well-equipped shop, but I will probably have only slightly better luck convincing readers to set up an efficient office than to write a business plan. My description (see pp. 21-27) of the 50-sq. ft. office from which you can briskly manage a million dollars a year of construction will probably be mostly skipped over. That is unfortunate. For whatever the scale, when you are a builder, you are in business, and you need a place where you can dispatch your phone work and paperwork comfortably, swiftly and accurately. You need a thoughtfully equipped office.

One of the first tasks you can perform in your office is to write a policy statement—part of the fourth month's work. In my experience, builders do not often compose a company policy until well into their careers. By writing it earlier, however, you can clarify your thinking on issues that a construction company faces daily, such as employee behavior at the job site. When the issues arise, you will be able to resolve them speedily instead of groping for a response. "Here in paragraph five," you can tell an employee, "just as you agreed when we hired you."

Among the most important tasks you will perform in your business are accounting and estimating. In my sample plan, I have allocated for each task two full months to set up a system. With help from the detailed material in subsequent chapters, you may find two months enough. But take more time if necessary. Without sound accounting systems, you cannot maintain financial control of your individual projects or of your business as a whole. And when an opportunity arises for a substantial project, you want to have in place an estimating system that will allow you to bid it accurately. A single bad bid can bring down the curtain on a construction business.

A good estimate often leads to a contract for a project. Therefore, in the sample business plan, development of an estimating procedure is followed in the ninth month by development of legal documents. You create a contract tailored to your own operation, and along with it a form for that crucial extension of a contract—the change order.

Like sound estimating, accounting and contracts, complete insurance distinguishes solid construction companies from the shakier competition. Buying insurance is usually easy enough, but buying it intelligently is another matter. In the tenth month, you should therefore investigate both worker's-compensation and liability insurance, then locate brokers who can provide good policies when you are ready to take on projects.

As you near the end of your plan, perform a few final steps in the 11th and 12th months: Prepare for and pass any contractor's licensing examination required in your state. Even if none is required, study the building codes, lien laws and other regulations governing builders. Get a business card. Don't spend a chunk of working capital on an expensive custom card, but spurn a generic card like the one that features your name in black ink flanked by a red hammer and framing square. A thousand other start-up builders who have gone into and right out of business have used that card. Get any necessary local business licenses. If your company will use a name other than your own, you will need to file a "fictitious name statement" at your city hall or county offices. Develop a phone file and set up a portable office (pp. 27-31).

Gerstel's Business Card

In keeping with a commitment to minimal overhead, a business card is the only advertising Gerstel uses. His card, of the 1,000-for-$25 variety available at copy shops, uses a standard typeface arranged in a pleasing vertical format.

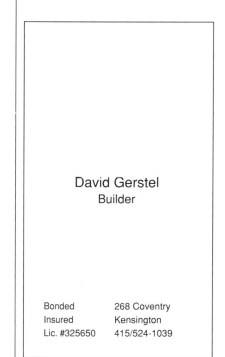

David Gerstel
Builder

Bonded   268 Coventry
Insured   Kensington
Lic. #325650   415/524-1039

If your plan parallels my sample plan and you can stick to it, by the beginning of the second year you will be ready to look for projects. If you enjoy a "hometown advantage," all you need do is post a notice at your church or community center, or mail out a letter introducing your new business to acquaintances. Quite likely, you will soon receive calls for your services. On pp. 76-81 I will discuss the whole gamut of other possibilities for finding work, from classified ads and cold-canvassing neighborhoods to soliciting at real-estate and architects' offices.

Once you begin contracting for projects, you have available several paths toward establishing a full-fledged company. You will find it useful to extend your business plan, laying out in detail the steps in the path you choose to take. One natural sequence:

• Do small side jobs while continuing as an employee in a larger company. Though it may feel like overkill, use your estimating, accounting and contract-writing systems for these projects. The fluency you develop now will serve you well later.

• Give up your job and work full time as an independent tradesperson, hiring temporary helpers and subcontractors as needed. Do small projects—decks, popouts, bath renovations—for which you can perform most of the construction yourself, even while staying on top of your paperwork and management duties.

• Continue doing small projects until you have full confidence in your estimating, accounting, contract-writing and project-management skills. If you love working with the tools too much to give them up, remain an independent tradesperson until you feel attracted to the management challenges of a bigger operation.

• If you do decide to organize a company, develop a crew. Do not hire a green apprentice first. Look for a lead, or someone who can develop into one. Extend lead responsibilities to him or her as quickly as practicable. Hire additional employees to work under your lead.

• Once you have a reliable crew teamed with a good lead, you can begin removing yourself from the job site. You have made yourself a new job running your construction company. Now you can concentrate full time on the management tasks we will examine in detail in the rest of this book.

A BUILDER'S TOOLS

Office
Portable Offices
Shop
Policy Statement
Working Capital
**Insurances: What, Why
and How**
Insurance: Liability
**Insurance: Worker's
Compensation**

OFFICE

As a small-volume builder, you will probably handle your office work at home. But you do not want to conduct business from the dining-room table or over the household phone. For the sake of your marriage, you don't want your spouse to field calls from angry subcontractors or clients. For the sake of your reputation, a prospective client's first call to you should not be answered by your gregarious preschooler or sullen adolescent. You need a place for your office and business phone that is as insulated as possible from family territory.

If you're lucky, you'll have a spare bedroom available for conversion into an office. If not, consider partitioning off a corner in the garage or basement and feeding in an electrical circuit. If that isn't possible, search for an unexploited space—a closet filled with abandoned toys or a plastered-over cavity beneath a stairway. In the worst case, you can create an office in a corner of your least-used room. I first ran my business from a walk-in closet and later operated out of a corner of our bedroom. Now that the kids have moved out, I enjoy the luxury of an ample worktable built against two walls of a sometimes guest room.

The important thing is that your office, wherever its location, should provide a comfortable space to perform the crucial work of evaluating prospective projects, estimating, bookkeeping and financial analysis. Otherwise, you will struggle to do these tasks or even neglect them altogether. If you simply do not have room for something better, you may have to work at a conventional desk shoved against the wall. I did for many years. But when I moved to my pre-

An Office in a Closet

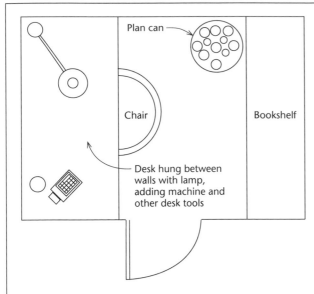

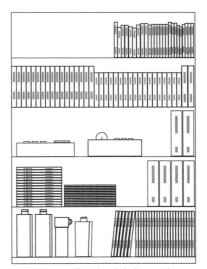

Floor-to-ceiling bookshelves with reference books, back copies of magazines, files, phone, answering machine and briefcases

When he first went into business, Gerstel set up his office in a walk-in closet. Though compact, it provided a private, quiet place to do estimating, bookkeeping, contract writing and other office work. The low office overhead (about $7 monthly) let Gerstel rapidly accumulate the working capital necessary to build his company.

sent setup, it was as if manacles fell from my wrists; the bicycle I had been pedaling uphill with the brakes on was suddenly coasting down the other side. I felt free. The paperwork that had been grueling became pleasurable.

The spacious worktable in my current office measures 12 ft. along the combined length of its two legs. Built of two solid-core birch-veneer door blanks trimmed with 1x2 redwood, it is supported by ledgers along the walls, by file cabinets at either end, and by a plywood cleat screwed to the underside where the door blanks meet. Because no supports protrude below, I can swing easily along the table's whole length. On the short leg I have space to open a set of working drawings when I am punching out a bid. The long leg holds equipment used in bidding and other paperwork. In fact, the table is organized as a miniature assembly line, with supplies and equipment positioned so that work can be processed efficiently from the "in" to the "out" end.

At the "in" end sits a plan can for drawings and specs awaiting bids. The top of a two-tier clamp-on in-basket holds the letters, bills and proposals that flow in each day. The bottom holds a "tickler" for

Gerstel's Current Office

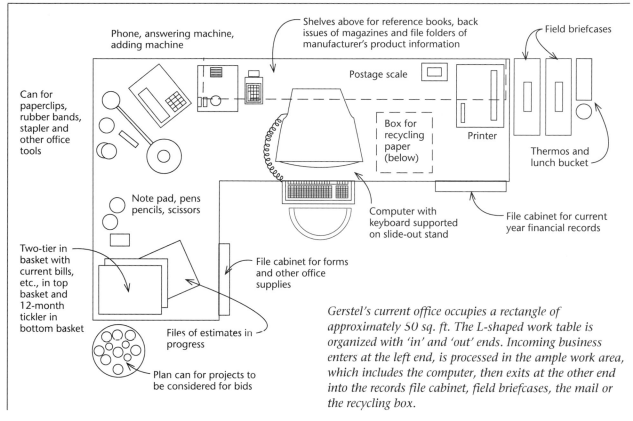

Phone, answering machine, adding machine

Shelves above for reference books, back issues of magazines and file folders of manufacturer's product information

Field briefcases

Postage scale

Can for paperclips, rubber bands, stapler and other office tools

Box for recycling paper (below)

Printer

Thermos and lunch bucket

Note pad, pens pencils, scissors

Computer with keyboard supported on slide-out stand

File cabinet for current year financial records

Two-tier in basket with current bills, etc., in top basket and 12-month tickler in bottom basket

File cabinet for forms and other office supplies

Files of estimates in progress

Plan can for projects to be considered for bids

Gerstel's current office occupies a rectangle of approximately 50 sq. ft. The L-shaped work table is organized with 'in' and 'out' ends. Incoming business enters at the left end, is processed in the ample work area, which includes the computer, then exits at the other end into the records file cabinet, field briefcases, the mail or the recycling box.

reminders of crucial future obligations, such as tax payments. You can purchase fancy tickler files from office-supply companies, but I use a stack of six manila folders. Each one holds the reminders for two months of the year. I rotate them so that current months are always on the top for easy checking during my weekly day in the office. Below the basket I keep the files of estimates in progress.

Like good kitchen cabinets, file cabinets should have full extension hardware that gives ready access to the stuff at the rear. My "in" cabinet stocks my standard forms, arranged in the order they're usually needed during the qualification, estimating and construction of a project. (These invaluable forms are illustrated and discussed in subsequent chapters.) On the surface of the worktable, compact storage canisters contain an array of paper clips, rubber bands, scissors, pencils and pens, a paper punch and a letter opener.

At the corner of the table are the first of several electronic gadgets. The phone, which has a different number from the household phone, features programmable dialing and automatic redial, two extremely useful options. The first saves many small increments of time, which add up over the year, by dialing your regular suppliers and subcontractors. The second helps you to break through the busy signal at the building department or lumberyard when you urgently need service.

Next comes the answering machine. I shudder at the thought of operating without one. Answering machines enormously increase efficiency. When you have questions for subs or clients, you record on their machines. They send their answers back to yours. You do not have to try repeatedly to get in touch personally (and waste time with the obligatory small talk when you do). When you are working in your office, to maintain concentration, you can screen calls—I take only the urgent ones and answer the others later in a single batch. Your subcontractors, as well as other repeat callers, will especially value the efficiency of your answering machine. Provide them with a short announcement—they don't want to hear a cut from the Rolling Stone's Greatest Hits or your kid's violin solo. On the other hand, they will appreciate ample time for their own messages. Thirty seconds is not enough; three minutes usually suffices.

Some builders believe an answering service makes a more "professional" impression than a machine. Perhaps that was the case a few years ago. But now even blue-chip legal firms use machines (or "voice mail," which is really just a higher-tech answering machine). Nobody mistakes an answering service for a personal secretary anymore, and often a service sounds a good deal less professional than a thoughtfully recorded announcement tape. Worse, service personnel can easily garble technical messages from architects, plumbers, electricians or other subs. An answering machine gets the message right.

Cabinetmaker Steve Nicholls's office features a high table where he can stand or work from a high stool. If you are putting in long office hours, you may welcome the opportunity to alternate between standing and sitting.

Answering machines are so cheap and helpful that you may want to keep a backup to make certain that you are never without one. The same holds for the calculator with paper tape, which is essential for estimating and bookkeeping. Without a tape, you punch numbers into your calculator twice, praying the two totals will agree. With a tape, you can quickly spot and correct any errors or omissions, and you also get a permanent record of the calculation for later reference.

Conspicuous by its absence from my worktable is a fax machine. I have questioned whether faxes are cost effective for a small-volume builder like myself and have elected to do without. Other builders make a convincing case that their faxes save them much time taking quotes or information from subs, suppliers and designers, and increase accuracy of communication. They also point out that the right fax can double as a copier, at least for small runs. They tell me that, as with a chopsaw and a computer, you don't appreciate a fax until you've used one. By the time you read this book, a good fax may be available for a few hundred dollars; if so, it may well pay for itself in a reasonable time.

The next item along my electronics row, the computer, raises such complex and controversial topics that I will introduce the items on the "out" end of my work table before discussing it. First there is a postage scale. Before I discovered I could purchase my own, I made a lot of unnecessary trips to the post office to weigh mail. Second is a recycling bin—better that trees remain as animal habitats, or at least be used for lumber, and that Sunday's paper be made from the wads of commercial mail that land in our offices every day. Third is a fireproof file cabinet, which protects the current year's financial records. If the office were to burn down, I would be able to recover the records and keep a handle on my business.

M y basic worktable with all the equipment, except the computer, cost me about $1,200. When I began writing this book, to have added a computer with software for word processing, estimating and accounting would have increased the cost of setting up an office by 600%. For that reason, I intended to advise new builders to refrain from buying computers until they had built up ample working capital, a much more important tool. In the intervening years, however, computer costs have plunged. Computers are still not cheap. And I still recommend that if you are just starting up and must make a choice, that you give priority to working capital (pp. 35-39). But with prices having dropped so sharply, you may now be able to enjoy the great benefits provided by a computer.

With a word-processing program such as Wordstar or Xywrite III, you can readily create, revise and produce documents that can sharpen the operation of your business. On my own computer hard disk, I currently maintain several dozen forms and other documents. When I need to amend a document, the computer allows me to do so with great ease. For example, when a bulletin from the contractors' license board informed me that builders were required to provide the board's address in their contracts, with two minutes of keypunching I was able to update my contract in my computer's memory. With similar efficiency, I constantly improve and create other documents. Whenever I need a copy, I simply turn on the computer and printer and generate one. For the frequently used forms I keep in my "in"

Documents Stored on Gerstel's Computer Disk

Gerstel uses his computer to create, improve and print the following documents and forms he uses in running his company:

Job-prospect form
Client references
Professional references
Price-planning agreement
Estimate "Do" list
Preliminary-estimate checklist
Estimate checklist
Specialized estimate checklists
Materials quote form
Subcontractor phone-quote form
Subcontractor quote checklist
Assumptions to accompany bid
Contract
Subcontract
Change-order form
Construction "Do" list
Safety checklist
Time card (basic)
Time card (for specialized projects)
Job-cost form
Payment schedule
Application for payment
Time-and-materials bill
Policy statement
Ad for carpenter
Ad for crew leader
Hiring interview checklist
Subcontractor interview checklist
Letterhead
Standard letters
Credit application

file cabinet, I can print out a whole batch—no more running to the copy shop to stock up on forms or replace outdated ones.

Ironically, though I am emphasizing document management, when builders hear "computer" they rarely think first, if at all, of forms. Instead they envision number crunching and crisp financial records, estimates and bids pouring from a printer. In fact, while the document-management advantages of computers are indisputable for the small-volume builder, applying computers to number work is a much more dubious proposition. Several times I have interviewed builders who had bought computers to help sort out snarled accounting procedures. But instead of achieving clarity, they had become even more lost, having combined their confused accounting methods with the perplexities of computer operation. Meanwhile, their businesses were hampered by lack of the working capital the computers had absorbed. Even for the well-organized small-volume builder, accounting by computer may be overkill. Michael Thomsett, author of *Builder's Guide to Accounting* (see Resources, p. 223), has told me that a builder who writes fewer than 100 checks a month will be less efficient with a computer than with manual accounting systems such as I recommend in Part Three of this book.

Estimating by computer is nearly as controversial. A program may contain bugs that skew your final numbers. If you do relatively few estimates, maintaining your program may cost you more time than it saves. The wrong estimating program can reduce your flexibility, forcing you to organize or present your estimates in ways at odds with your preferences. On the other hand, estimating by computer does save you the labor of punching up totals and proofs on your calculator. It can also speed calculation of the cost for many line items in your estimates. Your customers will appreciate the clearly printed summary of numbers the computer can provide. With high-quality estimating programs now reportedly on the market for a few hundred dollars (*Journal of Light Construction*, February/March 1990), you may find it worthwhile to try one out.

For all the pros and cons, one thing is virtually a given in the computer debate: You should adopt computer record keeping and estimating only after considerable experience with well-organized manual systems. In addition, you must for a time operate your manual systems in parallel with the computer to make sure it is giving accurate results.

"Fine," you may say. "Now which hardware and software should I buy?" I can't tell you that. Computers, printers and programs change so rapidly both in cost and capability that by the time you read these lines, any advice I write will be as antiquated as tips on how to operate a hand auger. I can only suggest that before you select a computer, you get an overview of current technology by reading the latest manuals and magazines or by attending introductory classes. Next, find a dealer who appears likely to stay in business, sells name-brand equipment and backs it with reliable service. Do not buy cheap. The information stored in your computer will soon greatly exceed the value of the hardware and software. It makes no more sense to store and manipulate that information with an unreliable machine than it does to purchase cherry crown molding and then miter it with a dime-store handsaw.

Similarly, to get full value from your computer, you should learn ten-finger typing. To use a computer with only pecking skills is silly. Within a matter of months, the extra hours required to peck out documents will exceed the time it would have taken you to learn ten-finger technique. If you do have the funds to install a computer and printer, expand the slot in your business plan (see p. 14) for creating your office from part of a month to two full months or more. Then you will have allowed yourself the time needed to master your machine.

PORTABLE OFFICES

With a well-organized and well-equipped worktable, you can move through your paperwork so briskly that you will need to spend relatively little time in your office. Most of the day you'll be calling on prospective customers, conferring with designers and supervising projects in the field. But there you will often find yourself in need of the same equipment and supplies stocked in the office. One builder solves the problem by taking his entire office with him. He has installed it in a recreational vehicle, which he drives to his job sites and appointments.

For a company the scale of my own, a pair of small "portable offices" does the trick. The first is a briefcase of the type called a catalog case. When I first brought my case to a job site, a jocular apprentice promptly nicknamed it "the office." I added "portable," and the name stuck. I always take it along when I go into the field. Its 1-cu. ft. interior contains sufficient files, forms, references and equipment to enable me to do much of my management work wherever I happen to be. Among the principal items in its three large pockets are:

- A field checkbook. For efficiency, I try to pay all bills from my home office in one weekly session. But occasionally a new supplier requires a check on delivery. Or a plumber, electrician or other sub who is pressed for cash wants the favor of payment immediately upon finishing a phase of work. With a checkbook in the portable office, I can always accommodate them.

- Licenses and insurance certificates. Among the most frustrating experiences in contracting is standing in line at the building department only to be denied a permit for lack of some license or certificate left behind at your office.
- Forms. I carry copies of necessary computer-generated forms and an ample supply of change orders.
- Calculator with paper tape. This is useful in the field for totaling costs and markup on larger change orders and also serves as a backup to the office machine.
- Phone flip file. Over the years, from conversations with other builders and at job sites around town, I have become acquainted with hundreds of subcontractors, suppliers and other construction professionals. Their names, along with notes on their reliability and the quality of their work, are logged in my flip file. Having the file at hand in the field is enormously useful. I am able to provide crew and clients with the names and numbers they need and to find numbers quickly myself when I must do phone work from a job site.

Gerstel's portable office contains tools for efficient field management.

- Project records. I keep records for each project under construction in a heavy-duty three-ring binder until the job is complete to the last details. The presence of the records goads me to see that those details are promptly dispatched. Nothing erodes a customer's appreciation so much as your lagging on the completion of a few items. A rave reference can decline to so-so because you install a showerhead and the last few cabinet pulls a month after completing the rest of the project. A customer who can't bathe and must claw open the cabinets will not be a happy camper.

For years I got along nicely with a single portable office. But as my company prospered, my first case became steadily more pregnant with project records and management tools. Finally, it gave birth to another "office." I pulled an old briefcase out of storage and shifted to it items I use for sales and field estimating:

- Prospective projects file. In recent years, I have had calls for far more work than my company can handle. But having gone through several recessions, I have learned not to take work for granted. I make a record of every possible project, so that when the company does need work, I can open the prospects file and find a couple of dozen potential jobs.

- Sales tools, including portfolios, copies of my reference list and other items (discussed on pp. 83-84), that I use in meetings with prospective customers.
- A compact hairbrush and shoe polish, so I can spruce up for a sales call after leaving a dusty job site.
- Field estimating tools, including a flashlight, tape measures, stud finder, binoculars (for checking conditions on steep roofs) and a camera. The photos help me remember site conditions when I'm back in my office estimating. They can also help answer any later questions about conditions existing before construction. Clients become hyperaware of their homes during construction, sometimes notice existing damage for the first time, and may wonder if your crew or subs caused it.

My portable sales office showed its value recently. I stopped off at a new store that provides its clients with cabinets and recommends contractors for kitchen remodels. I described my company's services to the store's owner, opening my sales case to give her a copy of our reference list and to show her my portfolio. Not long afterward, she asked us to remodel her own home. She had been impressed, she said, because of all the contractors who had solicited her business, I had been the only one who came equipped to show his company's capabilities.

When he calls on a prospective client, Gerstel likes to have on hand everything he needs to explain his company's services and to take the first steps in making an estimate.

My portable field office, on the other hand, does not customarily attract such compliments, but is more likely to provoke amusement. As first a project record, then a stapler, then the calculator, then the phone flip file comes out, a client will stare down into the case and demand, "What else you got in there? A canned ham?" But I find both portable offices enable me to make use of dead time, such as when waiting for subs or inspectors, to knock out phone calls, organize jobs or work on estimates. Some of the efficiencies the portable offices make possible—like putting a prospective sub's name directly into the phone file rather than making notes, then transferring them to the file back at the office—may seem tiny, but small efficiencies do add up. I learned this firsthand over two decades ago while working as a garbage collector in a small rural community. When my partner and I took over from the previous guys and followed their procedures, the route took us two hours. But over the course of a few months, we invented hundreds of new moves. Each saved only a few seconds, but in total they cut the job by 75%, to a total of about half an hour. In the same way, by looking for small efficiencies in my construction-management duties, I have pared the core work of running my company down to about 20 hours weekly. To be sure, that 20 hours does not include estimating. But I often do no estimates for months because we are booked far ahead by clients who value the attention we pay to efficient organization.

SHOP

Cabinetmakers need a shop. Millwrights need a shop. Hairdressers need a shop. Builders do not. As a builder, your shop is at the job site. You can probably not compete, either in cost or quality, at tasks like cabinetmaking and millwork, which really require a shop. The people who do those jobs all day and every day have the machinery and know-how to leave you in their dust. Unless you want to do their work for the sheer love of it and are prepared to subsidize it with the profits from other aspects of your operation, leave it to them.

I know small-volume builders with cavernous shops—really grand square footages beneath high ceilings, stocked with neatly arrayed equipment. I am often impressed at first glance and give in to the envious feeling that to operate from such a palace would be the height of construction glory. But on closer examination, I realize that the shops stand empty and the equipment is unused most of the time, gobbling money for rent or mortgage payments and maintenance and loading the builders down with unnecessary overhead and maintenance chores.

Even more suspect than a shop is a warehouse for all that tempting material left over at the end of projects. Construction consultant Walt Stoeppelwerth acutely defines a small-volume builder's warehouse as "a place where you put something and never take it out." You spend time and money to carry it there, keep it there and keep it clean. My own "shop"—really a compact storage area—is a 10-ft. sq. space in my garage. Its deep shelves accommodate most of the equipment for my crews. Usually, however, they stand empty. Our equipment is where we need it, at our project sites, installed on the nifty portable storage systems (pp. 205-206) we move from job to job. If we have a lull between

In lieu of a shop, Gerstel maintains a compact storage area in a 10-ft. by 10-ft. space in his garage. The deep shelves accommodate all company tools when they are out of use during slow periods.

projects and our equipment overflows the garage, I simply rent temporary storage space. (Or I offer to pay my carpenters to store the equipment in their garages and encourage them to use it for side work until the next company project begins.) Occasional temporary rentals burden the company with far less overhead than permanently rented or purchased space.

On my company's projects, leftover material—whether salvaged or new—is promptly moved along rather than being embalmed inside some expensive real estate. In order of preference, it is:

- Taken back to the supplier, ideally on the return trip of a delivery truck, for a refund
- Taken to the used-building-materials yard by an apprentice, who uses the proceeds to acquire a new tool
- Put at curbside with a sign "Free firewood, free building material"
- Taken to a recycling center
- Taken to the dump

You may have trouble resisting the allure of a warehouse. ("I think the warehouse mentality has psychological roots in the need to accumulate," laments the wife of one builder whose garage is stuffed with gleanings from his renovation jobs.) You may have even more trouble resisting a big shop and the latest in expensive machinery. The big shop seems like the real thing, the abode of the substantial citizen, the throne room of a construction baron. It resonates with a promise to make *you* a baron, while the horsepower of the machinery promises somehow to become your power. You may decide to go for it all. But go for it with a clear view. With each purchase, pause to consider if you are putting into place a needed building block for your company or if you are merely stroking your ego. Are you adding strength to your enterprise? Or are you engaging in personal consumption and disguising it from yourself as business investment?

POLICY STATEMENT

Years after I began general contracting, I realized that certain issues were coming up for discussion between myself and my crew again and again. To get the record straight between us once and for all, I wrote my first policy statement. Now I ask prospective new employees to read our policies and to let me know whether they can accept them before beginning work. If they cannot, they should go elsewhere. If I do get their approval, I take it as an encouraging sign that they will be comfortable with my authority. That is important, because as a small-volume builder you are at various times one of the gang, in a supportive role and the boss. Few employees have trouble accepting your friendship or support. But not infrequently you will encounter individuals who will feel compelled to undermine you at every opportunity. As one builder says, "They get to thinking I'm dad, and they're pissed off at dad." Meeting their jibes when you go to your project sites can distract you from your work and take the fun out of building.

Policy statements can cover an encyclopedia of issues. Here I will comment on a few of those most crucial to small-volume builders. As with other matters that are subject to state and federal laws, I must emphasize that I can only call issues to your attention and not provide resolutions you can adopt directly or verbatim. You (or your lawyer or other professional consultant) must write the actual policies you will use.

In my experience, the question of who provides which tools stirs up the most contention between contractor and crew. When I started out in the trades, no problem existed. Carpenters were expected to have a standard complement of hand tools, all of which fit in a single hand-carried box. Builders provided the power saws and drills and the utility items like cords and ladders. That is all the equipment we used on a daily basis, and costs were not serious for either carpenters or contractor. Now a radically different situation exists. Many new large and costly tools, like air-driven nailers and compound chopsaws, are standard issue at job sites. Hand tools have been electrified and gone up in price 1,000%. Builders find the cost of the larger equipment burden enough and hesitate also to provide all the power hand tools, especially since carpenters are likely to treat them with less care than they would if they owned them. On the other hand, purchase and maintenance of all the electric screw guns, planers, routers, biscuit joiners, sidewinders and sanders can take a chunk out of a carpenter's paycheck. The carpenters may enjoy acquiring these tools, but resent wearing them out on company projects.

To maintain morale, you need a clear, fair policy stating who provides what tools and how carpenters will be compensated for use of their own tools on company projects. You may find it useful to develop a tool list, stating the tools required for different levels of employees from apprentices to leads. You might reasonably expect apprentices to work the flea markets for hand tools and to refurbish them with pride and for little cost. Perhaps you will reward their dedication with a bonus of an occasional power tool. For the journey and lead levels, however, the full complement of equipment you require may be worth thousands of dollars. A more formal means of compensation becomes necessary. You can provide it in a variety of ways, as long as you make sure you are operating correctly with re-

Policy Items

Policy statements can become encyclopedic. Here is a partial list of the items typically included. Builders should write their own statements, then have them checked by an attorney who specializes in labor law.

Equal opportunity	Time clock or sign-in system	Attending seminars, meetings
Proof of citizenship	Pay period	Company parties
Physical exams for prospective employees	Payday	Causes for discipline
	Vacations	Grievance procedure
Probationary period	Paid holidays and show-up pay	Termination
Hiring of relatives	Sick call-in	Exit interviews
Work hours	Sick pay	Moonlighting
Lunch and breaks	Health insurance	Use of company equipment
Dress requirements	Worker's compensation	Noncompetition agreement
Safety practices	Unemployment insurance	Accepting gifts from clients, inspectors, etc.
Tool requirements	Disability insurance	
Confidentiality regarding company issues	Automobile insurance	Communication with clients
	Pension plan	Behavior in clients' home
Full-time status and benefits	Profit-sharing	Alcohol and drug use
Overtime occurrence	Tuition assistance	Theft
Overtime pay	Bonuses	Obscene and loud language
Penalties for lateness	Loans	Vehicle appearance
Severance pay	Leaves of absence:	Sexual harassment
Performance review	military	Site cleanup
Raises:	personal	Environmental protection
automatic	disability	Policy flexibility
merit	jury duty	

spect to both labor and tax law. Some builders pay a "tool allowance" on an hourly, daily or weekly basis. Others cover all tool maintenance, including replacement if tools are stolen from the company site or completely worn out.

Beyond tools, a number of other policy issues frequently arise:

- Safety. Go beyond encouraging safe practices at the job site. Require them, and the more specifically the better. Some workers pay little attention to safety. Explain to them that their recklessness endangers not only themselves but other workers as well. And, because injuries elevate costs—for crew turnover and worker's-compensation insurance—recklessness also threatens your company's financial health and its ability to compete for projects and provide jobs. Therefore, taking risks is not an employee's prerogative; rule it out by company policy.

- Discrimination or harassment on the basis of gender, race or religion. Rule it out in firm language.

- Standards of appearance. Itemize your expectations with respect both to personal and vehicle appearance at the job site.

- Breaks. Dave Matis and Jobe Toole, authors of the first-rate *Paint Contractor's Manual* (see Resources, p. 223) insist, "You just can't afford to let your employees set their own schedules." They require that lunch and breaks be taken at a fixed time. In my company, with highly motivated workers at the job site, I have found the reverse to be true. The crew takes responsibility for timing the breaks; I require only that they take the morning and afternoon breaks required by labor law. Likewise, your break policy may be determined by the type of people you employ.
- Evaluations and grievances. Employees want to know how they are doing, and they need opportunities to air their gripes. By scheduling regular evaluation and grievance sessions (and including them in your tickler), you ensure communication.
- Discipline. Clearly state the transgressions—such as theft of company tools, repeated tardiness, loafing or alcohol use at the work site— that will result in suspension or termination. Avoid levying fines or docking wages as punishment; you are not a court of law, but simply an employer, with power to make decisions regarding employment.
- Termination. A troubling subject. Some experts say your company policy should provide for either employer or employee to "terminate the relationship at will." However, a good building company is unified by a sense of obligation between builder and crew. You can undermine that sense by declaring to your employees that it would be formally correct for them to be dismissed without warning or to walk off without notice. Rather than sticking solely with that cold, even if legally correct, position, you may also wish to express that in case of termination, both employee and employer will do their best, within practical bounds, to take into consideration the needs of the other.

Builders should not unilaterally impose new policies on established employees. I've made this mistake, and it has stirred up much resentment among the crew. If you are drafting your first policy statement and already have a stable crew, give them a list of the issues you want to cover and ask for input before you start writing. Once you have a draft, request their opinions, and revise it as is indicated and reasonable.

In creating a policy statement, you have several options. You can write up policies in your own language. You can collect policy statements from other builders and from business magazines and books, then edit the material into a set of policies appropriate to your company. In either case, have your product reviewed and corrected by an attorney or possibly by a personnel expert versed in your state's and the federal labor law.

Alternatively, you can have a policy statement professionally written from the outset. A lawyer may be able to tailor a boiler-plate version to your needs for reasonable cost. Companies like Paychex, a national payroll-service company, offer similar services. They will provide you with a computer-generated statement modified to fit your business and update it regularly to conform to changes in the law, all for a moderate subscription fee.

By whatever means you create a policy statement, do not expect it to be legally unassailable. Like builders, the experts are likely to disagree on the fine points of their practice, as I learned in an amusing fashion when I touched on personnel policy in a magazine article. The publisher's legal experts got into a dither when they saw my suggestions and ordered the editor to substitute a new paragraph. Following publication, the editors received a lengthy letter from an independent personnel expert. He proposed yet another slant on the issue, warning that readers who followed the policies described in the new paragraph—the one produced by the publisher's experts—were likely to get into "legal hot water." So what is the small-volume builder to do in the face of conflicting advice from the experts? The best you can.

 # WORKING CAPITAL

If a tool is defined as a means for doing a job, then working capital—a hefty balance in your business bank account—is probably a builder's single most important tool. But hands down, it wins the title of "most under-rated tool." Without working capital, you run the risk of alienating your clients, as one builder learned after putting her profits into a lease for a large shop instead of building up a cushion in her bank account. As a result, she says, "I find myself worrying about money instead of the job. It sets a bad tone with our clients, because I have to worry after them, asking 'When can you pay me?'" With inadequate working capital, you also risk losing the confidence of your subs, suppliers and employees. If you are short of money and delay payments to subs and suppliers, they may decline to bid your jobs or even close your accounts. If you are late meeting payroll, your workers become insecure and put out feelers for other employment. Replacing any of your people will cost you time and money and grief.

To circumvent that cost, builders short of working capital often incur another cost by procuring a line of revolving credit at a bank.

Startlingly, they sometimes seem under the preposterous impression that by becoming borrowers they have entered the big time. You hear them speak portentously of their "relationship" with their bankers, as if they had just arranged a date with a movie star. In fact, the bank's high interest on the short-term borrowed funds adds to overhead, making it more difficult to compete for projects and to maintain a margin for risk and profit. The interest feels especially heavy when it lands on you in the middle of a project already in financial trouble. "I didn't have that in my bid either," groaned one builder about the interest on the cash he borrowed to see through a balky and underbid project.

A shortage of working capital causes more insidious problems as well. If you are short of money and also short of work, you may get very nervous. Hungry for cash to make payments on equipment, to cover rent for your shop, or simply to provide yourself and your employees with a wage, you may estimate too tightly in the hope of winning a project. When you get one that you have underbid, it takes you still deeper into the hole.

By contrast, contractors who build up working capital enjoy tremendous benefits:

- Sub support. Subs who are always paid promptly are inclined to reciprocate with prompt bids and good service.
- Employee confidence. Employees who know they can count on being paid don't scan the horizon for another job.
- Supplier benefits and support. Many suppliers offer discounts for the quick payment you are able to make when you have ample funds in your business account. If you do find yourself in a cash squeeze and unable to cover their bills, suppliers who know you typically pay promptly are more likely to carry you than to cancel your account.
- Bank support. When your checking-account balance is normally sizable, if you overdraw, your bank likely won't bounce the checks. Instead it will cover them and notify you to get funds in place. Rather than being a borrower from the bank, you are a lender to it, and the bank is eager to keep in your good graces.
- Bonding capacity. Bonding companies monitor the financial condition of the builders they back. As *The Basic Bond Book* (see Resources, p. 223) points out, bond companies know that "it takes money to start up a job, to carry a company over a period in which there might be a dispute with an owner, to pay overhead, and to finance slow receivables." They require that their clients be well capitalized.
- Cherry picking. If you are not desperate for cash, you can pick and choose projects. You can handle those architect-designed and

The High Road and the Low Road

Two contractors, X and O, started out working for a spec builder, then separately went into business for themselves when he retired. X set up a high-overhead operation. He took out a bank loan, built a spec house, completed it just as housing prices were making a huge run upward and enjoyed a large profit. He put none aside as working capital, but used all his profit as down payments on two more lots. He took two more loans and built two more spec houses. The upsurge in prices continued. X made more big profits, got more loans and built more houses. Still he put none of his profit aside. Instead, in keeping with his position of a "rising young entrepreneur about town," as he fancied himself, he borrowed heavily to buy new trucks and heavy power tools, and to build a large home for himself.

Meanwhile, O took the low-overhead road. He bought an old truck and equipped himself with used tools from flea markets and auctions. Instead of going to a bank for loans, he contracted with customers for small remodeling projects. His business grew steadily, and he was soon netting high five figures annually.

O saved his money, bought a small home, tightened his belt and paid off the mortgage in one year. Now, with virtually no personal bills, he began accumulating capital rapidly. X and other friends urged him to invest in the booming real-estate market. But he liked the freedom and peace of mind that came from being free of debt. He decided to stay on his own road to prosperity. Abruptly, the boom period ended and recession set in. X found himself with four spec houses he couldn't sell because mortgage rates had climbed over 20%. He mortgaged his own place to make the payments on his construction loans. But interest rates stayed up and the four spec houses did not sell. Finally, X ran out of money and the bank foreclosed on his spec houses, his home and his trucks. To scratch out a living, he took on small remodeling jobs. He was back where O had started when both X and O had gone on their own a few years earlier.

O would not have been aware of the recession if he had not read about it in the newspapers. His business continued strong, as the high cost of mortgages encouraged people to remodel instead of buying new homes. The high interest rates added large amounts of interest to his working capital, and at the same time drove real-estate prices down. O took advantage of the situation to buy an abandoned apartment house, now digging into his working capital to get a very low price by paying cash. He rebuilt the apartments, and found that the rents more than covered his family's living expenses. Not yet 40, he no longer had to work.

Since he loved to build, O continued running his construction company. Because he did not need income from his company to pay personal expenses, his capital built up rapidly. He soon had a quarter of a million dollars in the bank. Friends urged him to invest in a new real-estate boom. "No," said O. He was making plans to use his capital to build spec houses without having to use bank loans.

government-sponsored projects that provide for 10%, and sometimes more, of each payment to be retained by the owner until completion. On a $90,000 project, the $9,000 retained could include all your profit and a sizable portion of your overhead. But with ample working capital you can afford to wait for it.

Finally, working capital makes you happy. You don't worry about your bills, because you know you can cover them. You sleep well at night. You can relax during your coffee breaks. Working capital also enhances your status. Word gets around the construction community: You have your act together. Your company is not just a hollow shell. It is sound.

Once you conclude that working capital is a must, you need to know how much you require and how to acquire it. In figuring the amount, don't confuse or mingle working capital with the funds you should have on hand to see yourself and your family through six months or more of hard times. That money should be in a high-interest personal account. Working capital should be at the ready in your business account to facilitate the smooth operation of your company. The exact amount you strive for will depend on the niche you work in. One authority concerned with larger commercial construction companies suggests you maintain working capital equivalent to 10% of bonding capacity. The National Association of Home Builders prescribes 5% of gross annual receipts. In my experience, 5% of annual receipts is tight. I prefer 8% or even 10%. That is a lot of money; at half a million dollars annual gross, it is $40,000 to $50,000. At a million, it is $100,000. Cash. In the bank.

I suspect some readers will be startled at the recommendation for such a large cash store and will wonder how one can ever accumulate it. Unless you have rich relatives or other patrons who want to capitalize your company (a bad break, actually, since the gift will likely sap your determination and muddle your sense of business reality), there is only one way to build capital: Save it. You must gradually put it aside out of your earnings. Here is one scenario:

- To create your initial nut, while you are still an employee, put a percentage of each paycheck into your business account.
- When you begin to do side jobs, put a large percentage of the additional income into your business account. (Probably you will be investing the remainder in necessary office and job-site tools.)
- When you have completed all the steps in your business plan and have set out on your own, do not spend all your profits on tools, and certainly not on personal goodies. Deposit a percentage of the receipts for each project into your working-capital account. Build toward a goal. For example, if you anticipate doing $100,000 in construction your first years, build toward a $10,000 balance in you business account.

You will readily find rationalizations against the accumulation of working capital. Among the favorites is the boast, intended apparently to project an image of entrepreneurial boldness, "I have every penny in my business." Working capital *is* money in your business. It is money most astutely positioned in your business.

From every side, you will be urged and tempted to spend your capital—we are a culture devoted to consumption. But to ensure success as a builder, you must resist consumption and save. William Mitchell, in his *Contractor's Survival Manual* (see Resources, p. 223), puts the point with nice bluntness. "Ignore that urge to buy the latest luxury car. Salt that cash away. Someday you will need the money in reserve. If you can't stand the thought of driving an old truck and answering your own phone, turn directly to chapter four, which deals with bankruptcy, because you are going to need it."

INSURANCE: WHAT, WHY AND HOW

Builders must carry general liability and workers'-compensation insurance. Liability coverage is required both by good practice and by savvy clients and designers, while workers' compensation is required by law. By the time you read this book, you may also be legally obligated to provide health insurance for your employees (and if you wish to run a sound business, you should in any case do so as soon as you can afford the premium).

With or without health coverage, insurance is expensive. Many builders pay more in liability and workers'-compensation premiums than they do in state and federal taxes. Because the cost is so great, small-volume builders often try to forgo coverage. Before succumbing to such temptation, however, consider that although the checks you will write for liability and workers' comp feel large, when viewed as a percentage of gross receipts, insurance costs can appear reasonable. For example, in recent years my company's combined outlay for liability and workers'-compensation insurance has run about 4% of total dollar volume. Also bear in mind that even at sites run by conscientious and capable builders, severe property damage does occur and workers do get injured. Through insurance we share our responsibility for damage and injury and prevent accidents from mushrooming into personal disasters for our employees, our clients and ourselves.

Though insurance is the largest single investment most small-volume builders will make, they frequently buy it without a small fraction of the forethought given to the purchase of a plunge router. They learn nothing about the product they are purchasing. They learn little more about the salesperson. More than once, I've heard builders unhappy with their insurance agents rationalize their relationship: "Well, you know, he seemed like a nice guy, and his office was real convenient, just around the corner." So they stopped in, chatted briefly and made a purchase that would cost them as much as a Rolls Royce over the next half-dozen years.

If you choose instead to give serious attention to your purchase of insurance, your first decision will be whether to buy through an agent or a broker. Perhaps in your community you will find an agent with an excellent record of service to builders, who offers the policies you need. But purchasing from an agent can have a disadvantage. Agents represent only one company. They will, therefore, tend to be much more concerned with their company standing than with any individual client.

By contrast, brokers represent multiple companies. They need good relationships with these companies, but because they don't have all their eggs in one basket, they are freer to stand by their clients. Moreover, they can shop different companies on behalf of their clients, an impractical task for small-volume builders to undertake themselves. You do not have the time or expertise to evaluate the claim-settlement records of different companies. You cannot keep up with the constant changes in insurance-industry practices and in the structure of policies and coverages. A good broker, however, can effectively shop and evaluate policies for you, and recommend adjustments to suit changing conditions.

Whether you elect to go with an agent or broker, choose your person carefully. Insurance is complicated, and you need a competent representative as much as you do a capable attorney or accountant. In my work, in fact, I talk with my insurance broker much more frequently than with my accountant or attorney. To choose a broker:

- Ask builders you respect for names.
- Interview several brokers. Ask which policies they recommend and why. Ask them to explain trade-offs in cost and coverage. Evaluate the clarity with which they explain their products. You will often have questions, and a broker who can't give clear answers will not do you much good.
- Select one broker. Ask the broker for references, especially builders.
- Interview the references. Ask about the broker's availability. (My first broker, whom I picked much too casually, was hard to reach, and his staff did not understand my policies.) Ask if the broker initiates improvements in the policies. (My current broker, whom I selected by the method suggested here, has several times found ways to improve the coverage and lower the cost of my policies on his own initiative.) Ask about responsiveness to claims.

If the first broker on your list does not check out, move on to the second. You may settle on different brokers for liability and worker's-compensation policies. These two types of insurance are discussed in the following two sections.

INSURANCE: LIABILITY

As with other matters governed by federal and state law, my objectives in this chapter and the following one must be limited. I cannot give exact prescriptions for purchasing liability and worker's-compensation insurance. Because of the complexity of the subject, the density of the legal technicalities and state variations in insurance law, I will instead point out issues and questions to consider when purchasing your policies. You must select for yourself the coverage appropriate to your own situation.

The major liability policy you will acquire is known as CGL, for Comprehensive and/or Commercial General Liability. As the checklist at right suggests, CGL covers a wide range of risks. After reviewing a potential policy, you quite reasonably may wonder if there is some pocket of liability it does not cover. On this score, you may be able to put yourself at ease—I have been informed by an insurance attorney that for my state (California), if coverage is not specifically excluded from a policy, then it is automatically included. You will want to learn whether a similar ruling exists for your area.

Naturally, you should pin down any exclusions in your policy. If they are available as options and are germane to your operation, you may want to purchase them. One major coverage, sometimes excluded but available as an option, is "XCU," for claims arising from explosion; from collapse caused by excavation, shoring or underpinning; and from damage to underground utilities. Other exclusions, such as that for damage caused intentionally by yourself or your employees, may not be able to be removed from your policy for any price.

Another exclusion you likely will have to live with is the work-product exclusion. In technical insurance language, coverage is excluded for "that particular part of the property that must be replaced, repaired or restored because of faulty workmanship." In other words, with a work-product exclusion in your policy, you won't be covered for defective work produced by yourself or your crew. For example, if your lead neglects to flash a window, your insurance company will not cover the cost of proper reinstallation when the window leaks. If flooring cups because your carpenter laid it improperly, you will be responsible for the cost of replacement. If the cabinets you hung fall off a wall, you will be responsible for repairing and rehanging them.

Builders are sometimes astonished when they learn about the work product exclusion. "Why bother to have insurance?" exclaimed one irate contractor. There is a good answer. The work-product exclusion is likely to be limited. In layman's terms, while you may be responsible for the defective work itself, the resulting damage may be covered. For example, if the leak at an unflashed window damages

Liability Insurance Evaluation Checklist

Builders typically need to rely on the guidance of a good broker when purchasing liability insurance. This checklist covers policy features that you should review with your broker.

Coverages
Premises medical coverage liability
Contractor's protective liability
Extended broad form property
    damage liability
Blanket contractual liability
Employees as additional insureds
Spouses of partners as additional
    insureds
Damage to property
Personal injury and advertising
    injury liability
Nonowned watercraft liability
Incidental medical malpractice
    liability
Fire legal liability—real property
Supplemental payments
Exclusions
Collapse caused by excavation,
    shoring or underpinning
Damage to underground utilities
Intentional damage
Failure to build to specification
Design
Work product

Policy type
Claims-made
Occurrence
Tails

Limits
Per claim
Aggregate (annual) limits
Limitations on percentage of work
    subcontracted

plaster in a room below, your policy may cover the repair of the plaster, though it excludes the cost of properly reinstalling the window.

Another aspect of liability coverage builders find disconcerting concerns design. Here the problem is that coverage may be ambiguous, neither specifically included or excluded. Meanwhile, you may find it impossible to remain uninvolved in design. Even when you don't produce the overall design, you will often work with sketchy plans from architects or other designers, which force you to take responsibility for design decisions. For projects where you do receive detailed drawings, there is still a rather wide grey area where construction and design merge. In fact, a standard architect's contract holds builders responsible for construction "implied" by the drawings.

Opinions differ as to how far liability policies cover contractors for design. One view has it that while policies typically do not cover design work, they do not exclude it either. Therefore, the argument goes, insurance companies must settle claims arising out of design work. Unfortunately, even as they are settling a claim, an insurance company may cancel the policy because it does not want to cover a builder who does design.

If you do design work and your liability policy does not cover it, you may want to look for an "errors and omissions" policy, the type of coverage taken by architects. But be prepared for difficulties in your search. Errors and omissions policies are not widely available to contractors. When they are, the premium is likely to be high. (Indeed, errors and omissions coverage is so costly that small-volume architect's offices often do without it.) As a result, you may have no choice but to live with uncertain design coverage under your liability policy.

When you purchase liability insurance, you may have the choice of a claims-made or an occurrence policy. The distinction is so mind-numbingly technical that I hesitate to trouble readers with it. But if you do not understand it, you run the risk of paying enormous amounts of money for liability insurance and still ending up without the coverage you need.

With a claims-made policy, your current policy covers current claims. These policies have an advantage. Costs of settling claims are always rising due to inflation, but because current policies are written with regard for current costs, if a claim develops in connection with work done years ago, you should have enough coverage to settle it.

On the other hand, claims-made policies have a potential disadvantage. If you terminate your current policy (when you retire or change careers, for example), you may not have coverage if a claim develops in connection with past work. Even if you extend the policy by purchasing a "tail," you may be in jeopardy, because you may not be allowed to expand coverage sufficiently to keep up with inflation.

If you do purchase a claims-made policy, carefully determine with your broker or agent what tail coverage will be available should you terminate the policy. But the greater likelihood in today's market, I am informed by experts, is that as a small-volume builder you will be steered toward an "occurrence" policy. The advantages and disadvantages of occurrence insurance are roughly the reverse of claims-made. With occurrence insurance, according to the *Associated General Contractor's Guide to Construction Insurance,* (see Resources, p. 223), "The policy in effect at the time a loss occurs covers that loss irrespective of when a claim is brought." Thus, if you terminate an occurrence policy, it still applies to work you did while it was in effect. This means that if a claim is made in 1990 on work done in 1955, the occurrence policy in effect at that time applies.

On the down side, the coverage may not be nearly enough. As the *AGC Guide* explains, "Since current claims are paid from past policies, the limits of liability may be inadequate because they have not increased over time to reflect inflation." If you purchase an occurrence policy, discuss with your agent ways of maintaining adequate coverage should you drop the policy. One possibility is to set your limits of liability high enough to account for at least moderate future inflation.

The amount of liability coverage necessary is a matter of personal judgment. There are no simple formulas. In determining the amount, take into account the potentially high legal costs of settling claims, as well as the costs of any property damage or personal injury for which you might be responsible. Since legal costs can be a major portion of settling, you want to have in effect a policy likely to cover those costs if you do face a major claim. Thus, the fact that you are new in business and doing small projects on modest homes does not mean you can settle for minimum coverage. If you seriously damage one of those homes and injure the occupants, the legal costs of settlement could exhaust your minimal coverage, leaving you personally responsible for further defense, medical and repair costs.

In determining how much coverage to purchase, you must also determine whether your policy provides "per-claim" or "aggregate" limits. In the past, the limit was likely to be per claim. Now policies typically offer an annual aggregate limit. There's a big difference between the two. Suppose you have a limit of $500,000 dollars and are hit with a $410,000 claim early in the year. With a limit-per-claim policy, you still have $500,000 coverage for each additional claim. With an aggregate limit, however, you have only $90,000 remaining, which might be insufficient to handle any further claims filed during the remainder of the year.

Additional Liability Coverage

The following policies may be purchased in addition to a comprehensive general liability policy.

Business auto

Insurance companies require that a builder's work vehicle(s) be covered by a business policy. For business policies, vehicles are categorized by use—commercial, retail or service. Rates vary from one category to another. You can save yourself a substantial amount of money by making sure you are placed in the lowest applicable category.

Any auto or nonowned auto

Here you cover yourself against claims that involve someone else's vehicle used for company work. For example, a carpenter picks up lumber on the way to the job in his or her truck and has an accident. You are named in the ensuing lawsuit. With a nonowned auto endorsement on your business auto policy, you're covered. (The employee, however, is not covered under the nonowned endorsement, but must have an independent liability policy.)

Equipment theft

For a surprisingly modest additional sum, you may be able to obtain a rider that provides cash or even replacement-value coverage against tool theft.

Umbrellas

You may find it cost-effective to raise your limits of liability by buying an umbrella rather than by increasing the limits on your general liability policy. Umbrellas are tricky and have their pros and cons, so if you consider one, discuss it carefully with your agent or broker.

Course of construction (builder's risk)

If you build on spec, you need to insure your projects against such losses as fire damage during the course of construction. If you build under contract for clients, you should include in your contract a clause that requires them to provide builder's risk insurance, or to pay you to provide it.

Fortunately, increasing your level of coverage may not be terribly burdensome. For example, with my policy, $300,000 in coverage would cost me 4.6% of the wages for my crews. By raising coverage to $1,000,000, I incur only an additional 1.5% cost.

Once you have chosen a policy and its limits, your broker or agent will tell you your premium and payment schedule. The premium can be figured either as a percentage of projected gross receipts or of projected payroll. If it is based on payroll, you have an opportunity to save yourself money. Some policies for independent tradespersons and small-volume builders automatically include a fixed allowance (about $35,000 with the policy I carry) for your presumed hands-on work at job sites in the projected payroll amount. If you are truly a manager and not working on your projects as a carpenter, you may be able to get that allowance pulled from the projected payroll. My broker did, and reduced my annual premium $2,100.

INSURANCE: WORKER'S COMPENSATION

By law, as an employer, you are financially responsible when an employee is injured on one of your projects. You must provide medical care, compensation for lost wages and rehabilitation. The benefits are due on a no-fault basis. In other words, you must provide them whether or not you are responsible for the injury; employees are entitled to the benefits even when they are at fault. If your carpenters fall off a scaffolding because you did not provide a safety rail, they get their benefits. If they fall because they took the rail off, they still get their benefits. In return, (at least in theory though apparently in diminishing fact), you are protected from being sued by injured employees.

You can best cover your obligation to employees by purchasing a worker's-compensation insurance policy. (Self-insuring is not practical, even where legal, for small-volume builders.) The policy should include employers' liability coverage for those instances when an employee does manage to sue you over a work-related episode. In some states, worker's compensation with employers' liability insurance is available only through a government controlled fund, and in others only from private carriers. In some states, both options are available.

As a no-fault system, worker's compensation is intended to support injured workers while at the same time controlling costs. In recent years, however, the system has been hard hit on the cost side. Medical and rehabilitative care has become increasingly sophisticated and expensive. With rising frequency, employees sue their employers, thereby costing the insurance carriers heavily for defense. The carriers' increased medical and legal burdens have in turn driven up their administrative costs. All three costs—medical, legal and administrative—contribute to the steep rates for worker's compensation. For example, as shown in the payroll report above, the base rate for carpenters in California in the late 1980s exceeded 20%. If you were paying the base rate (and in a few paragraphs I will describe how that rate can be modified upward or downward), a $20-an-hour lead carpenter cost you an additional $4 an hour in worker's-compensation premium. In a busy year, insurance for that carpenter adds about $7,500 to your job costs.

Because of the high cost, some contractors devise schemes for circumventing a part or even all of the compensation premiums on

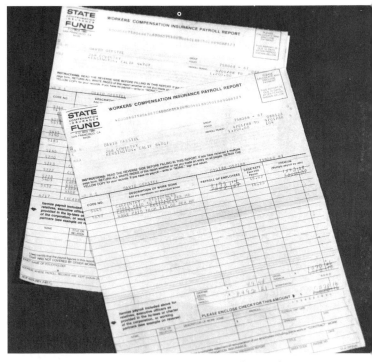

Many insurance companies, such as the California State Compensation Insurance Fund, whose payroll report form is shown here, will allow a breakdown of wages by trade, and it can result in a large savings.

their workers' wages. Unfortunately, by doing so, they endanger themselves and their employees and also rip off fellow contractors. In the event of serious injuries, they can face financial ruin in a lawsuit. Their injured employees may not get the medical care that will get them back on their feet and may be condemned to lives of poverty and misery. Meanwhile, of course, the cheating contractors' portion of the worker's-compensation burden is picked up by those builders who do pay their fair share.

There are far better ways than cheating for builders to keep their worker's-compensation costs at reasonable levels. Legitimate tactics include:

- Purchasing from a carrier who allows you to break your employees' wages into different trades, since some carry lower rates
- Purchasing insurance from a carrier with a record for paying high dividends
- Maintaining safety at your job sites
- Taking care of workers who get hurt

Some worker's-compensation carriers will allow you to break the wages of individual employees into different trades. Because different trades pay very different insurance rates, you can realize a substantial savings. For example, in my company, the crew, who are basically carpenters, spend about one-fifth of their time at trades other than carpentry. The base rates for those trades average about 10%, as opposed to 20% for carpentry. Therefore, by reporting wages under several trades, I lower my overall worker's-compensation costs by about one-tenth. You may feel that such a savings is not worth the trouble of breaking down and reporting wages by trades. However, with a well-organized time card and a computerized payroll service such as I describe in Part Three, the breakdown happens automatically as a byproduct of other tasks.

Not all worker's-compensation carriers offer the possibility of a "dividend"—redistribution of excess premium to the policy holders. Even those that do cannot promise a dividend or forecast its amount. You can, however, use reports of past dividend levels to project likely future dividends. For example, I switched carriers when my insurance broker offered a compensation policy from a carrier with a recent history of far higher dividend payments than I had been receiving. In my first year with the new carrier, my dividend more than doubled, to about 40% of my premium. In other words, on a $15,000 premium, I recovered about $6,000.

The best way to control worker's-compensation insurance costs is by careful attention to safety at the job site. Your safety record dramatically affects your dividend. For example, with the dividend described above, if my crew had incurred a single accident requiring medical treatment, my dividend would have dropped from 40% to 32% of the premium.

Even more critical, your safety record determines a so-called "experience modification factor," which is applied to your base rate. If your experience is good, if your crew works safely and sustains only few and mild injuries or none at all, your base rates can be modified sharply downward. If your experience is poor, your rates can soar. Builders with very good safety records can see their rates cut by a third or more. Builders with poor safety records can see the rate rise by several hundred percent. As a result, in a state like California, with its 20% base rate for carpenters, builders with good safety records might actually pay out only 10% to 12% of carpentry wages for worker's-compensation insurance. In addition, they may recover a large portion of the premium as a dividend. As a result, for a $20-an-hour lead carpenter, they may pay only $1.50 an hour in worker's compensation. Meanwhile, builders with poor safety records could see their rates zoom to 60% of wages and collect no dividend. Their effective insurance rate on the same $20-an-hour lead could be $12 an hour, destroying their ability to compete.

If, despite your best efforts, workers are injured, give their care high priority. See that they receive immediate medical attention. Even small injuries should receive treatment. Otherwise, they can worsen and result in greater disability for the employee, larger benefit pay-outs for the insurer and an increased "experience modification" for the employer. Seriously hurt workers need special attention. Often they will be badly frightened, and not only by the injury itself. They may be facing long-term unemployment or even the complete loss of ability to practice their trade. Often they do not understand the extensive benefits available to them under the worker's-compensation laws. If you are not on hand to support them, to explain their benefits and to speed those benefits along by prompt reporting of the accident to your carrier, an injured worker can feel abandoned. He or she may then go outside the compensation system and seek help from an attorney. Though the worker may well not see any increase in benefits as a result of litigation, the cost of settling the claim, and with it your future compensation rates, can soar.

As the rates go up, down goes the financial viability of your company. Crew morale plummets. If you run a small company and one member is hurt and then neglected, other workers will be alienated. If instead you support the injured worker, company morale and performance are strengthened even as worker's-compensation costs are controlled.

PART 3

KEEPING THE BOOKS

Why and Who

From Pile System to
  Pegboard and Spreadsheet

Essentials: Petty Cash
  and Taxes

Adding a Crew

Prospering: Job Costing
  and Receivables Journal

Consolidation and
  Sophistication

WHY AND WHO

You may view bookkeeping as a distasteful chore forced upon you by the Internal Revenue Service. But in fact the numbers for your tax returns are only one important spin-off from your financial records. Even if there were no IRS, you would need good books to manage your company; businesses in this country kept books long before income taxes existed. Good financial records spotlight the low and high places in your business's performance and warn you of potential twists and turns in the road ahead.

Builders let bookkeeping slide for two main reasons. First, they see it as a distraction from the "real work" of bidding and running projects. One builder told me, "For a long while I was so busy out in the field 'making money,' I didn't take time to keep accurate books and make sure I really was making money." When he finally did analyze his records, he found that his remodeling projects were subsidizing a $3,000-a-month loss by his cabinet shop. (Since he preferred cabinet-making, he closed his remodeling operation to focus on the shop, and with the help of detailed bookkeeping turned it around so that it now earns him a good living.)

Second, builders are apt to interpret a hefty checkbook balance as evidence of financial success. They don't realize that the balance (and more) may be spoken for by subs' and suppliers' bills that have not yet arrived in the mail. Regularly, I learn of builders who thought they were in good shape because they had cash in the bank, but who actually were staying afloat on deposits from new jobs, which they used to pay bills from previous jobs. When business

Essential Accounting Goals and Tasks for Small-Volume Companies

To know where you stand financially during the year and with each project, you must rigorously track income and costs. Tracking income is easier—make sure you know what you are owed, then collect it, bank it and record the deposits. Recording costs is more challenging. You must track them by date and in a large number of categories and also by project. If you are thorough in recording income and costs, you will be able to accomplish the five crucial tasks at the bottom of the chart.

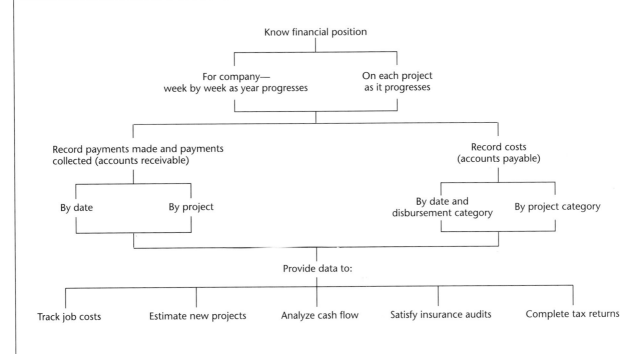

slumped and money stopped coming in, they could not pay the bills piling up on their desks.

With well-organized books, instead of focusing on only your current bank balance, you will be able to evaluate your company's overall financial performance. You'll know which categories of expenses are running high and where you're underinvesting. You'll be able to track each project as it proceeds and distinguish the kinds of projects that earn your living from the types you're giving away. You'll see which phases of work your crews perform efficiently and where you must improve.

When you set up a good bookkeeping system and begin to analyze financial performance, your mind-set naturally evolves from a tradesperson's to that of a company manager. Tradespeople work for an hourly wage and hope for enough hours to make a decent living. Managers must conceive of earnings not as an hourly return to themselves, but as a return to the company over months, quarters and even years. Tradespeople would never work for nothing an hour. Managers might operate at break-even, or even accept a small loss for a few weeks, to preserve an opportunity for good earnings down the road.

But who should do the actual work of bookkeeping? As your business matures, you may find it cost effective to hire a bookkeeper. In the early stages of organizing your company however, I recommend you do your own bookkeeping—with occasional input from your accountant. Write checks, make deposits, fill out ledgers and journals and complete your own federal and state tax returns. Repeated close exposure to your numbers will give you a vivid understanding of your company's financial and tax affairs, which in turn will enhance your moment-by-moment management of your company and its projects.

You may resist the hard work of learning about and doing your own books. If so, consider this: Of a building company's assets, cash is the most important. It is also the easiest to steal. If you do not understand your finances and entrust them to the wrong person, he or she can fleece you clean. And you will never even know it.

From Pile System to Pegboard and Spreadsheet

Some years ago I received from the Internal Revenue Service a copy of that dreaded form letter, the one that invites you in for an audit. I met with my accountant, mapped strategy and reviewed my returns for any weak spots, then drove to the appointment rehearsing my defense plans: Be courteous. Answer questions concisely. Hand over documentation briskly. Volunteer nothing (name and serial number only, sir). I felt ready, until I took a seat in the waiting room. There I was hit with premonitions of huge bills for taxes unpaid, plus late penalties, plus charges for...for what? Negligence! Incompetence! The other people in the waiting room looked like they were about to be taken by green death, and they were armed with impressive cloth-bound ledgers and steel file boxes. If these heavily equipped folks were as vulnerable as they obviously felt, what sort of fat pigeon must I be? My only ammunition was an old shoebox packed with stacks of receipts and checks stored in recycled manila envelopes.

My turn came. I was shown to a small cubicle, one of dozens on a large, open floor. But instead of the predatory character I had feared, I was greeted by a slender man suited in bland beige who gave the distinct impression he would rather have been fishing—a speckled trout leapt from the green stream of necktie flowing down his shirt front. He questioned me methodically in a remote voice. Three-quarters of an hour into the audit, precisely at noon, he declared it over with a judgment of "no charge." No charges for unpaid taxes. None

Shoebox Bookkeeping System

The most basic record-keeping system is so simple that it fits in a shoebox. It includes a business checkbook and a set of envelopes for storing receipts from projects and for general expenditures.

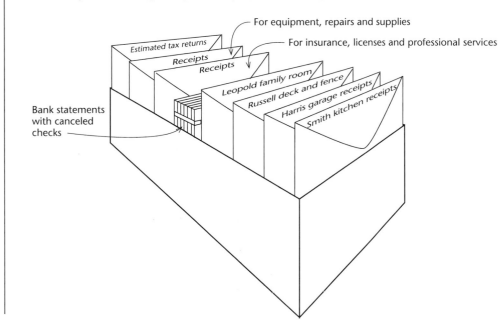

for improper books. On the contrary, as I packed up to go, the agent complimented me on the orderliness of my records.

The shoebox bookkeeping that served me so well at the audit is known in accounting lingo as the "pile system." In using it you simply sort invoices, statements, bank deposit slips and checks into piles. From them you can extract the numbers you need for tasks such as figuring earnings as the year progresses, tracking job costs and filling out tax returns. For example, to determine whether a project's costs are running as estimated, you pull the pile of project invoices from their envelope, total the invoices by categories such as materials or subcontractors, and compare the totals to your estimate. Similarly, to figure expenses for your tax return, you go through your pile of checks, totaling them by the categories specified on the IRS forms— office supplies, phone, postage, project materials, insurance and so on.

As my audit experience suggests, the pile system is not without value. It might get you by during the independent-tradesperson period of your business. Even then, and more so as you begin to hire employees and to take on larger projects involving multiple subcontractors, you will benefit enormously from stepping up to a pegboard (often also called a "one-write") system. A gem of a tool, the pegboard will astonish you with its low cost, simplicity and clarity.

One-Write Bookkeeping System

This drawing shows how the parts of a pegboard system go together. The journal and the checks are perforated to fit over the small stainless-steel pegs. The pegs, together with the clamp that snaps over them, hold the journal and the checks in exact alignment on the binder.

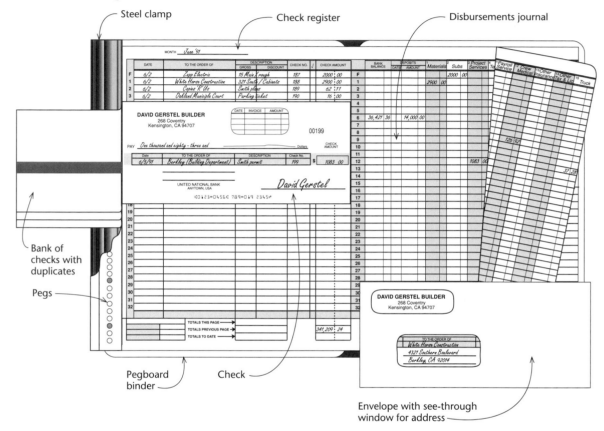

A pegboard system, such as the one shown above, with enough journals, checks and envelopes to last a year, should cost you approximately $100. Systems are available from many sources, including mail-order houses, but choose your supplier with the same care and criteria you apply to other professional help. With good bookkeeping so crucial to the prosperity of your business, the tools you use for it must be backed by good service. When I purchased my pegboard system, I interviewed several suppliers, finally choosing the one who came to my office rather than expecting me to come to his, and who explained his products most clearly.

Using a pegboard is delightfully easy. To pay a bill, you fold back the bank of checks until you have exposed the top one, then fill it out. The information you write onto the shaded line of the check carbons through to the duplicate and also to the check register.

(Thus, the name "one-write"—with one writing you get three entries.) You remove the check and duplicate by tearing along a perforated line. The check goes into an envelope with the address showing through the window. The duplicate is stapled to the invoice as part of your costs record. To record deposits, you open a line in the register by moving the bank of checks down a peg.

After making an entry in the register, you make a corresponding entry in the disbursements journal. ("One write" does not mean that you can actually make all the entries you need in a single writing.) The disbursements journal is really just a spreadsheet, which is nothing more than a piece of paper ruled into a grid of columns and lines. Across the grid you spread and group, then total, your numbers. The numbers are then readily available for your financial analysis.

How you organize your spreadsheet depends on the nature and needs of your business. Here I will describe my disbursements journal as an example of one way that works (see pp. 54-55). My disbursements journal, together with my entire bookkeeping system, is organized on a "cash" basis, meaning that income and expenses are recorded when the cash comes in and when it goes out. Cash accounting is generally considered acceptable for small-volume construction businesses. On p. 74 I will discuss the more sophisticated "accrual" basis. Your accountant can advise at what point in your company's growth it is necessary to switch from a cash basis to an accrual basis.

My journal's first column records my bank balance. The second records income. The next 34 columns record expenses in six groups: direct job costs, general expenses (also known as "fixed overhead"), profit sharing, capital investments, transfers from the company account for personal use, and adjustments.

The direct job costs—the costs incurred at our projects—are recorded in 11 columns and include:

- Materials. Costs for those materials that become a permanent part of a project, such as lumber, hardware and windows.
- Subs. Payments to electricians, plumbers, tilesetters and so on.
- Project services. Building permits and inspections, engineering and recycling.
- Temps. Payments to the temporary labor agencies that provide us with workers during demolition and excavation and occasionally for more skilled tasks. These costs were once included with subcontractors, but I wanted a clear look at the volume of business we do with the temporary agencies, so I gave them a separate column.

Pegboard Spreadsheet

The spreadsheet Gerstel uses with his pegboard system serves to record both income and costs.

	DATE	TO THE ORDER OF	DESCRIPTION		CHECK NUMBER	✓	CHECK AMOUNT			BANK BALANCE	✓	DEPOSITS		materials	SUBS	PROJECT SERVICES	TEMPS	NET WAGES
			GROSS	DISCOUNT								DATE	AMOUNT					
1	6/2/91	ZAPP ELECTRIC	15 MAIN / ROUGH		187		2000 00		1						2000 00			
2	6/2/91	WHITE HORSE CONSTRUCTION	321 SOUTH / CABINETS		188		2900 00		2					2900 00				
3	6/2/91	COPIES 'R' US	SMITH PLANS		189		62 11		3									
4	6/2/91	OAKLAND MUNICIPAL COURT	PARKING TICKET		190		16 00		4									
5	6/2/91	BERKELEY (BUILDING DEPT.)	SMITH PERMIT		191		1083 00		5							1083 00		
6	6/2/91	DEPOSITS: (2) JONES (CHK 129) 8,000 / WILLIAMS (CHK 494) 6,000							6	36,421 36		14,000 00						
7									7									
8									8									
9									9									
10									10									
11									11									
12									12									
13									13									
14									14									
15									15									
16									16									
17									17									
29									29									
30									30									
31									31									
32									32									

CHECK PROOF (IF USING GROSS & DISCOUNT)
COL. A $
LESS COL. B $
MUST EQUAL C $

TOTALS THIS PAGE →
TOTALS PREVIOUS PAGE →
TOTALS TO DATE →

(A) (B)

(C) CHECK AMOUNT
341,209 24

TOTALS THIS PAGE →
TOTALS PREVIOUS PAGE → 364,795 16 60,732 72 114,169 49 6702 80 426 01 5202 1
TOTALS TO DATE →

DISTRIBUTION PROOF: (IF USING DISCOUNTS)
CHECK AMOUNT (COL C) + DISCOUNT (COL B) = COLUMNS 1 THROUGH 23

DISTRIBUTION PROOF: (IF NOT USING DISCOUNTS)
CHECK AMOUNT (COL C) = COLUMNS 1 THROUGH 23

PLEASE ORDER FROM LOCAL MCBEE OFFICE. IF UNKNOWN, CALL 1-800-526-1272

PLACE ON TOP PEG   MONTH 6/91   DISBURSEMENTS JOURNAL — DIRECT J

The following seven columns, while still part of the direct-job-cost group, are also part of a crucial subgroup, namely, costs for the labor of our permanent crew. Labor costs typically are the most difficult for a builder to control and must be recorded and watched carefully.

• Net wages. The crew's paycheck amounts after tax deductions.

• Employee taxes, including Federal Withholding (FWT), Social Security (FICA), State Withholding (SWT) and State Disability (SDBL). (Note that adding net wages and employee taxes yields the crews' gross wages.)

• Employer taxes. The company's portion of FICA, as well as Federal Unemployment Tax (FUTA) and State Unemployment Insurance (SUI).

• Worker's compensation. Payments for worker's-compensation insurance are considered a direct job cost, because they are charged as a percentage of wages (pp. 45-47).

• Liability insurance. These premiums are also a direct job cost, because they, too, are charged as a percentage of wages (p. 41-44).

• Crew supplies. Here I record expenditures for supplies or materials that are used on the job but don't become a permanent part of it (in contrast to the items recorded under "Materials"). Examples are pencils, utility-knife blades, drill bits, power cords, and parts and

	8 WORKER'S COMPENSATION INSURANCE		9 LIABILITY INSURANCE	10 CREW SUPPLIES	11 PAYROLL SERVICE	12 CREW MEDICAL	13 OTHER INSURANCE	14 OTHER TAX AND LICENSE	15 TRUCK
EMPLOYEE TAXES		LABOR COSTS / OVERHEAD							
		1							
		2							
		3							
		4							
		5							
		6							
		7							
		8							

OVERHEAD, CONTINUED — PLACE THIS HOLE ON THE TOP PEG OF THE FOLDING BOOKKEEPER. — TRANSFERS TO PERSONAL ACCOUNT

OTHER REPAIRS	BANK CHARGES	POSTAGE	DUES AND PUBLICATIONS	LEGAL AND PROFESSIONAL	BOOKEEPING	OFFICE SUPPLIES	MISCELLANEOUS ACCT.	AMOUNT		TELEPHONE	TRAVEL	ENTERTAIN	PROMOTION		PROFIT SHARING	CAPITAL INVESTMENT		NON-DEDUCTIBLE	INTEREST TRANSFER	ADJUSTMENTS			GENERAL LEDGER DESCRIPTION	AMOUNT
						62 11	NOT	1									1							
							IN	2									2							
								3									3							
							USE	4		16 00							4							
7182 19	1420 00			308 12	548 74	692																		

labor for repairing power tools. In my experience, crew supplies run about 10% of gross wages, a substantial expense.

- Payroll service. Monthly payments to the payroll service that produces our paychecks and associated paperwork.

The preceding five columns mark the beginning of a most important subject—"Overhead." In accounting language, these five columns comprise what is known as "labor burden," or "variable overhead," because they burden you and vary in proportion to the amount of wages you pay. Along with variable overhead, small-volume builders incur substantial fixed overhead, general expenses over and above the costs incurred at your project sites. Farther along (pp. 132-138) I will discuss procedures for evaluating and charging for overhead in a bid, but to lay the groundwork, you must accurately track overhead on your spreadsheet. For my work, I record fixed overhead costs in 15 columns, including:

- Crew medical. I pay for employee medical plans even during occasional slow periods. Therefore, the expense is fixed, incurred every month.
- Other insurance. Payments for insurance, notably automotive liability and tool theft, having a fixed premium.

- Other tax and licenses. Annual taxes and licenses, such as city business taxes and my contractor's license.
- Truck. Operating expenses, including gas, oil and maintenance.
- Other repairs. Repairs, to office equipment for example, not included under crew supplies or truck.

The remaining overhead—bank charges, postage, dues and publications, legal and professional fees, bookkeeping, office supplies, telephone, travel, entertainment and promotion—are self-explanatory. Some of the costs I classify as fixed overhead actually vary with my company's volume of work. (For example, during busy periods we write more checks and make more deposits, which increases bookkeeping costs.) But the variation is so small, that it is appropriate and simpler to treat these costs as part of fixed, rather than variable, overhead. In your company, however, it might be necessary to group overhead costs differently. Overhead is tricky, so much so that major corporations have terminated profitable lines of production and kept losers going just because they did not properly account for overhead. For small-volume builders, the way in which overhead is allocated and analyzed must be tailored to the individual operation.

The next item in my disbursements journal—profit sharing—constitutes a group in itself, both because it does not fit into any other and because it is so important. I wish I could recall the name of the man who told me that "Profit sharing is a feature of any intelligently organized company," so that I could credit him. Profit sharing gives your employees a real stake in the company. Your long-term employees deserve the stake, and it can increase their loyalty and productivity.

Capital investment—expenditure for costly equipment, such as table saws and compressors—also is set off in its own column, partly because it requires special treatment at tax time (see pp. 60-61).

The next group, consisting of two columns, records disbursements that are not company expenses:

- Nondeductible. Here I record transfers to my personal account, which are not deductible for tax purposes. (For a corporation, such "draws" would be a company expense, since they would be issued as a salary or bonus to myself as a corporate employee. But for a sole proprietorship, all company income is considered the proprietor's, and it is not reduced by the amount taken for personal use.)
- Interest transfer. The working capital I keep in my company business account earns substantial interest. The bank deposits it automatically in my business account each month. But since I own the working capital, the interest is not really business income, and I transfer it to my personal account.

Disbursement Categories

A simple list helps you and your bookkeeper keep straight on which column to use for each expense.

Materials: Payments for all material put into projects (lumber, cabinets, countertops, etc.)

Subs: Payments to all subcontractors who work on projects

Project services: Designers, engineers, building permits and inspections, dumpsters and sanitary cans

Temporary labor: Workers from temporary agencies

Net wages: Net wages to workers

Payroll tax (employee): Taxes withheld from employee wages (FICA, federal income, state income, SDBL)

Payroll tax (employer): Taxes on wages paid by the company (FICA, FUTA).

Worker's-compensation insurance: Premiums

Liability insurance: Premiums

Crew supplies and repairs: All material and supplies used by the crew, which do not become a permanent part of projects — pencils, string, tape, protective paper and plastic sheeting, etc.; small tools (up to $200) purchased for use by the crew; all repairs to tools used by the crew

Payroll service: All payments to payroll service

Crew medical: Premiums

Other insurance: Vehicle insurance, tool insurance

Tax and license: Tax and license fees paid to city, county and state governments

Truck expenses: Gas, parts, repairs

Other repairs: Repairs not included under "Crew supplies" or "Truck expenses"; repairs to computer and other office equipment

Bank charges: Checking-account fee

Postage: Stamps, delivery charges

Dues and publications: Builders' group dues, seminar fees, subscriptions

Legal and professional fees: Lawyer fees, accountant fees

Bookkeeping: Payments to the bookkeeper

Office supplies: Paper items, tape, pens, pencils, paper clips, desk tools

Telephone: Monthly charges

Travel: Bridge tolls, parking tickets

Entertainment: Meals for customers

Promotion: Portfolios, photos, business cards

Profit sharing: Contributions to employees' IRA accounts

Capital investment: Equipment (all tools over $200); major vehicle overhaul

Nondeductible: Funds transferred out of business account to personal account

Interest transfer: Interest on business funds deposited in personal account

Adjustments: Corrections for errors

A final column, "Adjustments," is used for correcting any bookkeeping errors that may occur.

Note that my disbursements journal doesn't have an "Other" column. So far in my building career, I haven't had a need for one — all costs fit into one of the existing columns. Though your operation might require an "Other" column, if you find yourself recording many expenses under it, your spreadsheet needs further development. Expenses bunched together as "Other" provide no information useful for management or financial analysis. The column is just a pile, a throwback to shoebox bookkeeping.

For a disbursement journal to be useful, expense entries must be made consistently. You don't want to enter the cost of a repair to a power saw under "Crew supplies" one time, "Project services" the next and "Repairs" the next. Such inconsistency is a major source of trouble in many companies, since the skewed books lead to faulty analysis and conclusions. Avoid the problem by compiling and using a chart of accounts such as that shown above.

After completing each page of your disbursements journal, run an easy three-step proof to ensure that you've entered all numbers correctly. First, using your calculator with paper tape, total the entries in the register for the checks you have written. Second, total each disbursement column and then combine the column totals into a single total. Third, compare the totals of the check entries and the columns. If they are not identical, you have erred either in the register or disbursement columns. You can easily spot any addition error by looking at your calculator tape. If you made an incorrect entry, you'll find it by scanning your register and journal. With a quick adjustment, your books will be accurate and up to date.

ESSENTIALS: PETTY CASH AND TAXES

Whether you stay with a pile system for your accounting, adopt an off-the-shelf spreadsheet from your stationery store or go all the way to a pegboard system, you will need to track petty-cash expenditures. To keep them to a minimum, you can open charge accounts with your suppliers. Because my company maintains accounts with several dozen suppliers, in 1989 petty-cash layout was only a few hundred dollars, or about .04% of total receipts. With such small volume, I can handle petty cash informally. With each cash purchase I get a handwritten receipt marked "Cash payment," which I put in an envelope in my in-box. At year's end, I total the cash receipts, enter a single petty-cash sum in the check register and then enter each expenditure in the appropriate column.

A company with larger petty-cash expenditures needs a more formal arrangement:

1. Cash a check for several hundred dollars and post it to the nondeductible column of the disbursements journal.
2. Divide the money into envelopes, writing the amount on the face of each. Keep one for yourself. Give one to each crew leader, instructing them to replace money spent with the receipts.
3. When an envelope is down to a few dollars, remove the receipts, write "Petty cash" and the total spent in the check register, and enter the expenditures in the disbursements journal.
4. Add up the receipts. The total should match the amount of cash missing from the envelope. If not, post any amount lost to a "Cash lost" column.
5. Replenish the envelope with cash and repeat the cycle.

I recommend minimizing cash outlays by using charge accounts and checks for purchases whenever possible, because cash does get lost, and cash receipts offer less conclusive proof of a purchase than does a canceled check. Using checks also provides an opportunity to proof your bookkeeping. Each month, when the bank returns a statement and your canceled checks, reconcile their records with your own by balancing your checkbook. Most bank statements have easy-to-follow balancing instructions on the back; the book *Small Time Operator* (see Resources, p. 223) also gives an extremely clear description of how to balance your statement. If there's a mistake in somebody's bookkeeping, balancing will help you find it. In the event of an audit, the IRS likes to see canceled checks backing up your disbursements journal; they prove you really did make the purchases, not just pick up receipts from the sidewalk. And though the primary reason you keep good books is to aid in the management of your company, you also need the data for tax returns.

When you become self-employed, there is no boss to deduct weekly tax payments from your paycheck and deposit them with the IRS. You are required instead to estimate your own taxes and make payments—in April, June and September of the current year, and January of the subsequent year. The IRS forms include instructions for figuring the amount you should pay. Under 1989 law, the estimated payments must equal either your previous year's taxes or 90% of your obligation for the current year. If you underpay, you'll incur a stiff penalty. (Check with your state for its estimated tax requirements. They will likely parallel the federal.)

After the close of the year (between January 1 and April 15), you must file additional IRS forms to report your income, expenses and remaining tax obligation (or overpayment). For all the complaints we hear about the complexity of IRS forms, they are really quite logical and certainly among the simpler systems you deal with in the construction business. As a sole proprietor, your year-end tax return typically consists of a cover sheet, Form 1040 (the standard "long form," used if you itemize deductions) supported by two "schedules." On the schedules you total your income and expenses and perform preliminary tax calculations. The totals and preliminary calculations are then transferred to Form 1040, where you figure your final tax obligation or refund. (If you decide to incorporate, consult your accountant on corporate tax forms and schedules.)

After you pay estimated taxes the first time, the IRS will automatically send you four quarterly payment forms each year. Along with the forms come tables for calculating your estimated tax. Be careful not to underestimate—you can end up with a huge payment at the end of the year (one contractor had to mortgage her house to pay it), as well as stiff fines.

After Schedule C is completed, the number on its bottom line is transferred to Schedule SE, where Social-Security tax is figured. Then the results from Schedules C and SE are put on Form 1040, which is the basic 'long form.'

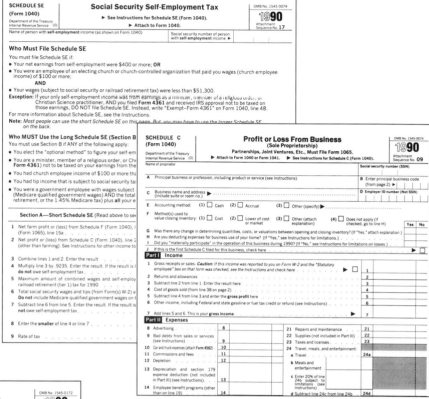

Form 4562 is necessary only if you make large investments in equipment.

The two schedules you need to support Form 1040 are:

- Schedule C. Here you report your gross income and expenses, and figure your net income or loss. Note that Schedule C has fewer and different lines than a well-developed disbursements journal has columns for costs. Several columns of the spreadsheet, such as totals for crew supplies and materials, combine into a single line on Schedule C.

- Schedule SE. Here you figure your Social-Security tax. (As a self-employed person, you have no employer to match your contribution, so you are responsible for the entire amount of Social-Security tax owed.)

Along with the schedules, you may need to include a Form 4562 for deducting capital expenses. At least under 1990 law, you need to file this form only if you buy a lot of expensive equipment.

Small tools can be deducted in the year of purchase. Larger and more costly items, such as table saws, compressors and trucks, though considered capital expenses, can also be deducted in the year of purchase as long as their costs total under $10,000. When you exceed that limit, you need Form 4562 to deduct, or in accounting language, "depreciate," the remaining capital investment across several years.

For example, during the first three quarters of a tax year you purchase a table saw, a nailing rig, a new engine for your truck and several large tools for $10,000. In the fourth quarter, you purchase a concrete vibrator for $950. The $10,000 can all be deducted in the year you spent it, but the cost of the vibrator must be depreciated on Form 4562. You will find that under the IRS's "accelerated cost-recovery system" (ACRS), you can deduct more of the $950 you spent on the vibrator during the first year and less later on.

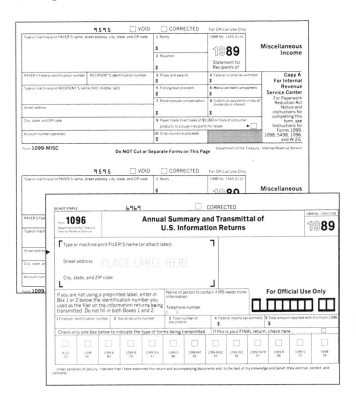

Form 1099 is used to report payments to some subcontractors, and Form 1096 accompanies and summarizes the 1099s.

One crucial point to note about the deductions you take on Form 4562: You do not make the deductions from the tax you owe. You make them from your income, then pay taxes on the remaining amount. Inexperienced business people frequently misunderstand. They say, "I'll write it off my taxes," meaning I'll buy a table saw and take the $800 I spend on it off the taxes I pay the government (so I'm really getting the table saw free). No way. The cost of the saw comes off your taxable income, not off the taxes themselves. The savings would vary with your tax bracket, but at most it would amount to a few hundred dollars, not the full $800.

At year's end, along with Form 1040 and supporting schedules, the IRS requires you to file Form 1099 to report payments you have made to people other than employees. Fortunately, you are not required to file 1099s on corporations or on people to whom you paid less than $600 in a year, so the paperwork is not terribly onerous. (I can usually fill out my 1099s in a couple of hours.) There are several types of 1099s. Builders typically use the 1099-Miscellaneous form. You can obtain your 1099s, along with the 1096 summary sheet, from the IRS by phone or mail.

Before mailing your tax returns, photocopy them for your file. The IRS requires you to keep copies of the returns and all supporting records for several years, with the exact length of time depending on conditions, such as whether or not you have committed fraud. (Yes indeed, if you have committed fraud you must preserve the incriminating documents longer than documentation for honest returns.) Eventually, you may want to get rid of the receipts and checks. But never toss the returns. They provide a useful financial history of your company. You will find it fascinating after 15 years or so in business to survey your returns and see that you have grown from an independent tradesperson netting $13,000 to the head of a company that does a million dollars a year of construction.

ADDING A CREW

Hiring employees may seem a huge hurdle. But with good technique (see pp. 168-176), you can find the right people, and they can hugely increase the satisfaction you get from construction. My employees have taught me new skills, increased my income and positioned me to take on much more challenging projects. When you do take the step of adding a crew, you are moving from independent artisan to company manager. Paralleling that move will come a major increase in your bookkeeping responsibilities—payroll.

Because of the complexity and cost of payroll, builders often try to circumvent it by paying employees as subcontractors. It is so temptingly easy to write your workers a standard check rather than to go to the trouble of withholding and depositing their taxes and filing quarterly and annual payroll reports. It is tempting, too, to shuck the burden of Social-Security taxes and unemployment insurance, which, as a legitimate employer, you are required to provide.

Avoiding payroll burdens by categorizing employees as subcontractors can, however, cost far more than it saves. If the IRS discovers your gambit, they can eat you right up. For starters, they can demand all unpaid back employer taxes—Social Security, unemployment insurance and disability. Next, they dig into you for interest (compounded) on those taxes. After these appetizers comes the rump steak. If your employees simply cashed their checks and did not report their income, you can be asked to pay all their unpaid income tax, and interest on those taxes, too, for dessert. When the IRS is done, the state tax agency puts what is left of you on its plate.

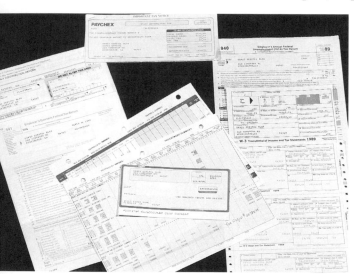

Gerstel's payroll service provides not only paychecks but also all tax documentation for the federal and state tax agency:
Biweekly: *Paychecks for employees, payroll journal summarizing wages, tax breakdowns by worker's-compensation-insurance categories, tax deductions*
Quarterly: *Federal tax return (Form 941), state quarterly report*
Annually: *Year-end federal forms and returns (W-2s, Form W-3 and 940-FUTA)*
As necessary: *Notifications of tax payments due*

Gerstel's Time Card

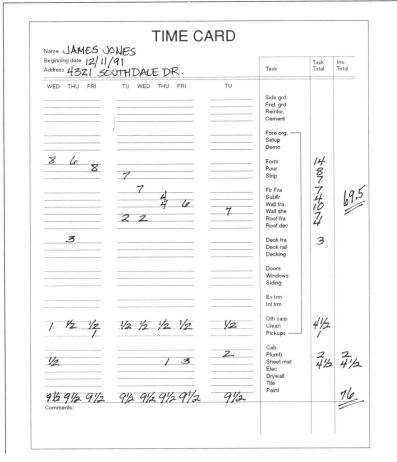

Gerstel's time card serves several functions. It records the total hours an employee works each day. It spreads the hours across various tasks for use in job costing. It allows the hours to be totaled both by tasks and worker's-compensation-insurance categories. Note that the card contains a built-in proof to catch mistakes. The hours for each day, totaled across the bottom of the card, must equal the hours for the insurance categories, totaled down the side. The time card starts with a Wednesday and shows only 8 days because Gerstel's company operates on a four-day week, and because pay periods start on Wednesday and run for two weeks.

In the long run, maintaining a legitimate payroll can be cheaper than trying to circumvent the tax collectors. These days, for about $50 a month, you can engage a computerized payroll service that will reduce your payroll work to a quarter of an hour every couple of weeks.

Engaging a payroll service will prove cost effective even when you hire only one employee. In contrast to the hands-on education you get from initially handling your own bookkeeping and tax returns, you gain little from doing payroll. (If you wish to give it a try, obtain the circulars needed for figuring payroll with a call to the IRS and your state tax agency.) You already have most of the knowledge you need about payroll if you took the trouble to understand the stubs on the checks you received as an employee. The rest you can get quickly from this chapter. From actually doing payroll, you will likely get nothing additional except a dose of tedium.

Your bank may offer a payroll service, but you may get more for less from a company specializing in payroll—look in the Yellow

Pages or ask other builders for recommendations. Whether you sub-scribe to a service or decide to grind out payroll at your desk, you will need to make three preliminary arrangements:

- Have each new employee fill out a W-4 form. (You can order W-4s from the IRS.)
- Create time cards to record your employees' hours.
- Make sure that your bank serves as a federal depository and can accept your deposits of employee withholding and other payroll taxes. If it can't, open an account at a bank that can.

By paying your employees biweekly instead of weekly, you cut your payroll expenses in half. At the end of each pay period, if you use a payroll service, you simply gather your crew's time cards, check for mistakes and call the hours into your representative. Some payroll services will deliver your employees' checks by courier service to your office within 24 hours. (Your bank should process the checks produced by payroll services as readily as they will their own.) With the checks come summaries of the wages, tax breakdowns by worker's-compensation categories, tax deductions and a notice to deposit any taxes due. At a later date, if you need detailed information about a payroll item, you can go back to these summaries. Therefore, when entering the payroll in your disbursements journal, you don't need to enter the data for each check individually. Instead, you can simply put the total of all the checks on a single line.

The frequency with which you must deposit payroll taxes is deter-mined by a complex formula, but basically, frequency increases as volume of payroll increases. When it is time to make a deposit, get it in immediately. The IRS jumps on delinquency in payroll-tax pay-ments and reports much more quickly than delinquent income-tax payments. If you neglect to file an income-tax return, the IRS may not query you for months. If you are late with payroll payments or returns, they will fine you or even impound your assets, and fast. I was once fined $300 because a payroll deposit was recorded at my deposi-tory a single day late, and I was able to get the fine rescinded only after insistent correspondence. Some payroll services will make the de-posits for you, and if they are late, assume responsibility for any fines.

At the end of each quarter, you must file payroll reports, or "re-turns," with the IRS. (Check with your state for its requirements, which will likely parallel the federal.) At year's end, you must file yet another set of reports, including Form 940, the FUTA return. As of this writing, federal unemployment tax is .8%, and you pay it only on the first $7,000 of each employee's earnings. As a small-volume builder, your annual FUTA obligation will likely be quite modest. For my own company, in 1989 it amounted to a few hundred dollars. State unemployment can be a different matter. Like worker's compensation, it can be "experienced-based" (pp. 46-47), and can run from a fraction of a percent all the way to 10%. Therefore, running a good company proves again to be good business, for a well-run company provides stable employment.

At year's end, you must provide employees with W-2 forms summarizing their income and withheld taxes for the year, and you must file a W-3, which summarizes the W-2s. (You will probably have to file a similar return with your state agency.) Filling out all these reports is even duller than making out the biweekly paychecks for your crew. If you have been grinding them out yourself, using a payroll service for the first time will be to encounter the relief of shade and a flowing well after a trek across a desert of boredom. The service will provide you with all the paperwork you need file, completely filled out and ready for your signature.

PROSPERITY: JOB COSTING AND RECEIVABLES JOURNAL

With a pegboard system, disbursements spreadsheet and payroll service, you have created the core of a sound bookkeeping system. You're probably well ahead of the great majority of small-volume builders. To go further and root your business in that select group of construction firms that prosper, you need one additional accounting procedure on both the cost and income sides.

On the cost side, although your disbursements journal provides a picture of overall company income and expenses as the year progresses, it does not keep you up to date on labor, materials and other expenses on individual projects as they move forward. And it does not give you final costs on the phases of work, such as framing, installing windows or running interior trim, within those projects. To retrieve that information, you must do job costing.

Job costing provides essential information both during and at the end of a project. As you build, it tells you whether you are within budget. If not, you can look for ways to tighten up while still maintaining quality. You might find your crew is going beyond the scope of the plans without alerting you to the need to write change orders. Maybe that new, and talkative, apprentice is distracting your seasoned workers. Perhaps material is being stolen. So you bear down on the change orders, instruct the apprentice to quiet down and focus on the work, and ask your lead to secure the material overnight and on weekends.

Job-Cost Card for a Remodel Project

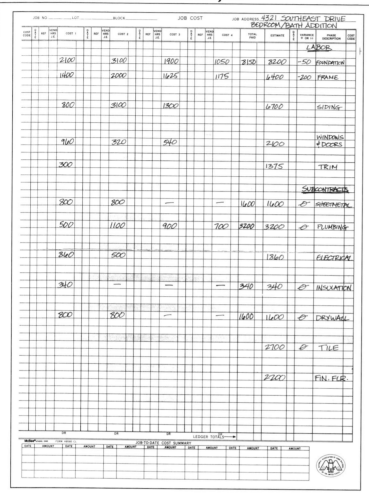

COST CODE	D E	REF	VEND HRS J.E.	COST 1	D E	REF	VEND HRS J.E.	COST 2	D E	REF	VEND HRS J.E.	COST 3	D E	REF	VEND HRS J.E.	COST 4	TOTAL PAID	ESTIMATE	D E	VARIANCE + OR —	PHASE DESCRIPTION	COST CODE
																					LABOR	
				2100				3100				1900				1050	8150	8200		-50	FOUNDATION	
				1400				2000				1625				1175		6400		-200	FRAME	
				800				3100				1300						6700			SIDING	
				960				320				540						2100			WINDOWS & DOORS	
				300														1375			TRIM	
																					SUBCONTRACTS	
				800				800				—				—	1600	1600		0	SHEETMETAL	
				500				1100				900				700	3200	3200		0	PLUMBING	
				860				500										1860			ELECTRICAL	
				340				—				—				—	340	340		0	INSULATION	
				800				800				—				—	1600	1600		0	DRYWALL	
																		2700		0	TILE	
																		2200			FIN. FLR.	

JOB NO _____ LOT _____ BLOCK _____   JOB COST   JOB ADDRESS 4321 SOUTHEAST DRIVE   BEDROOM/BATH ADDITION

DR   DR   DR   LEDGER TOTALS →   DR

McBee® FORM H8560 CL

JOB-TO-DATE COST SUMMARY

DATE	AMOUNT	DATE	AMOUNT	DATE	AMOUNT	DATE	AMOUNT	DATE	AMOUNT	DATE	AMOUNT	DATE	AMOUNT

Costing labor is most critical because labor is the cost most likely to get out of hand. On the project costed here, Gerstel is doing well. Several phases of work have been completed, and to date costs are within budget on the others. The job-cost card shown here is part of a McBee One-Write System, but you can easily make up your own cards.

Complete job-cost records for finished projects aid you in steering your company in financially productive directions. You learn which types of projects are profitable for your company and which are not. For example, one builder saw that he consistently failed to make a profit on the projects sent to him by kitchen designers. The reason was apparent: The designers' plans were so skimpy his project manager and leads were spending unforeseen hours figuring out details. The builder decided to drop kitchens in favor of more structural renovation, on which he was doing well. Meanwhile, a competing builder across town saw that one of his crews consistently preserved his profit margin on kitchen work. The reason: They liked building kitchens. The builder began assigning all his kitchen work to that crew and solicited more from the designers.

Knowledge of not only the final overall cost of a project, but also of the costs for phases within the project can help you select the right projects for your company. For example, you observe in your cost records that when building additions your company steadily earns your planned markup. You also note, however, that your costs have run 10% below your estimates for the demolition, concrete and trim phases while running 20% to 25% higher than expected for the framing. You are doing well on additions overall only because the over- and under-estimates cancel each other out. In an attempt to redress the discrepancies, examine both your estimating and site production. But meanwhile, don't make the mistake of setting yourself up for a loss by bidding a project that is heavily weighted toward framing.

As with every other essential business procedure, builders concoct all kinds of high-toned excuses for avoiding job costing. Mine once was that I did not want to know how much a job cost until it was completed, because if I were running over budget, I might be tempted to cut corners. To my good fortune, I hit a job that shocked me out of my rationalization without wiping me out financially. After completing a six-month project, I discovered I had made barely enough to cover my out-of-pocket costs and pay myself a modest salary. For the rest of my life that huge project was going to be sitting out there, and I would be responsible for it without a cent of profit against the risk. I was assuming the liability of an entrepreneur but settling for the pay of an employee, and, most ironically, I'd had an alternative. I was still in the phase of my career where I felt badly about imposing change orders—extra charges for extra work (see pp. 160-164)—on my clients. I had given away one extra after another. Had I tracked my costs, I would have realized I was not making a fair return on the job, and I would have charged for the extras. Now, except on the smallest projects, I track job costs as I build.

Like any other construction task, job costing requires systematization. As a first step, you need to develop the right documents for costing both labor and material. We have already discussed the labor documents, namely time cards, on p. 63. From the time cards you can easily work up the dollar figures needed for your job-cost card. For example, a carpenter costs you $20 in wages and 50% in labor burden for a total of $30 an hour, and spends 12 hours running joists during a pay period. Therefore, $360 ($30/hr. x 12 hr.) goes onto the job-cost-card line for framing.

To cost materials, you must obtain from your suppliers properly detailed invoices. You need separate invoices for each project; picking apart an invoice to separate out the costs for different projects is tedious and time consuming. You also need separate invoices for materials that go directly into a project and for those charged as either variable or fixed overhead, which would not go onto your cost cards.

Unit-Cost Record

```
                    UNIT-COST RECORD

Project:      JONES RESIDENCE

Phase of work:  REPLACE PERIMETER FOUNDATION

Dates of construction: 3-22-91 — 5-6-91

Crew leader:  D. LASSMAN

Other crew:  JOE BRASS / 2 LABORERS FROM TEMP AGENCY

Client:  JACK AND SARAH JONES

Description and scope of work:

  • REPLACE FOUNDATION FOR 2-STORY HOUSE

  • UNIT COST INCLUDES: SHORING OF HOUSE, REMOVAL
    OF EXISTING FOUNDATION, EXCAVATE & FORM
    FOR SPREAD FOOTING AND STEM

  • TOTAL LENGTH OF FOUNDATION: 147 FEET


Unit cost:  2.65 CREW HOURS PER FOOT LABOR

(CREW: CREW LEADER, CARPENTER & 2 TEMP LABORERS)

Conditions for work:

  ACCESS FROM STREET: EASY; LEVEL-IN WITH
    20 FT SETBACK

  ACCESS UNDER STRUCTURE: HORRIBLE; 14" to 16"
    CLEARANCE WITH 2X6 JOISTS — PROBABLY
    INCREASED LABOR BY 25% OR MORE OVER
    STANDING HEIGHT CLEARANCE.
```

Unit-cost records, such as the one shown here for a foundation, help Gerstel estimate costs for new projects. The written comments, made at the time of the project, allow him to modify old costs to fit the conditions under which new projects will be done.

f all project costs, labor is typically the most difficult to control. Material runs a distant second, with work contracted to subs coming in third. To gain control of labor costs and to be able to estimate them accurately for prospective new projects, you must take job costing a step further. You must, to use the industry term, record "unit costs" for as many phases and items of work as you can. Hanging a door is a unit. Miter casing a door is a unit. Butt casing is a slightly different unit. Installing a double-hung Andersen window or hanging a flush overlay laminate wall cabinet is a unit. One hour to miter-case both sides of an interior door is a unit cost; 1½ hours to hang a laminate wall cabinet is another unit cost. You must determine which unit costs will be useful to you in estimating and, therefore, which ones to calculate and record.

Whatever your decision, you must develop unit costs for yourself. You cannot rely on other people's figures. The publishers of construction-cost guides—compilations of unit costs—will tell you differently. One publisher claims that a cost guide can make a good initial price book, since it is allegedly easier to modify the guide's figures from your own experience than to build your own unit-cost record from scratch. It may be easier, but estimates built from the guide may be so inaccurate that they are worthless or, at worst, dangerously low. The data in publishing books are based on nationwide averages, but your costs are likely to be shaped by the requirements of your region and niche, by the quality of your work and by the expectations of your clients. I've compared my own costs to the published national averages. The spread so greatly exceeds my margin for overhead and profit that if I used guide figures I'd quickly go broke. One highly respected builder in my area owns almost every cost guide on the market. "None of them," he says, "is worth a damn, except to let you see how you are doing in a general sort of way compared to the rest of the nation." His opinion is shared by every experienced small-volume builder with whom I have ever discussed the subject.

In developing unit costs, you can travel one of several routes. Most easily, but also least useful for the long run, you can record them in dollars and cents right off your job-cost cards. For example, your job card shows $1,500 for forming the foundation. Your plans show 50 linear feet of one-story T-wall foundation. Therefore, your unit cost for forming one-story foundations is $30 per foot.

The problem with unit costs recorded in dollars and cents is that they need constant adjustment for inflation. To increase longevity of unit costs and avoid inflation adjustments, you can record the costs in hours. Thus, in the example above, your unit cost for forming would be 1 hour per foot.

For my cost records, I go a step further. Along with the hours used for a task, I record the conditions under which it was performed. For an estimate, I can then compare the past to the anticipated conditions and figure accordingly, rather than just plugging in a unit cost stripped of the detail under which it occurred.

The sooner you start recording unit costs, the sooner you will enjoy the payoff in vastly increased estimating speed and accuracy. You may very well find that after only a few projects you have the basic file of unit costs you need to estimate most of your work. Over the course of subsequent years and many projects, you can develop a comprehensive set of costs. However, just as with other accounting tasks, limit yourself to developing the unit costs that will really help you run your company. If you try to do too much unit costing, you may end up doing none of value. One conscientious young builder I know tried to figure unit costs for every phase of work on every project. He fell so far behind that he had no costs in his record relevant to the performance of his rapidly growing company. In my work, I have found it effective to pull just several labor unit costs out of each project. I concentrate on those for which I have only a few entries in my file or for which the entries are old and perhaps not applicable to my crew's current performance.

Receivables Journal Page

1990		DEBIT	CREDIT	BALANCE
				54643 00
12/4	CK. 103 DEPOSIT		1000 00	55643 00
1991				
4/19	CK. 107		10093 00	45550 00
4/19	CK. 108		5700 00	39850 00
4/30	CK. 109		3800 00	36050 00
5/13	CK. 110		6650 00	29400 00
5/18	CK. 111		2850 00	26550 00
6/06	CHANGE ORDER	4040 24		30590 24
6/15	CK. 112		5700 00	
6/15	CK. 113		4040 24	20850 00
7/01	CK. 114		6650 00	14200 00
7/07	CK. 115		5700 00	8500 00
7/23	CK. 10		5700 00	2800 00
7/22	CHANGE ORDER	5964 00		
7/27	CK. UNNUMBERED		5964 00	2800 00
8/01	CHANGE ORDER	189 00		2989 00
8/13	CK. 102		989 00	2000 00
9/10	CK. 3452		500 00	1500 00
10/01	CK. 3462		500 00	1000 00
11/09	CK. 3497		500 00	500 00
12/07	CK. 3521		500 00	

Gerstel collects from each of his clients at the job site and checks off payments on their contract-payment schedule. A receivables journal, kept at Gerstel's office, provides an essential check on the accuracy of his field collections.

The most conscientious attention to the cost side of your accounting will do little good if you neglect the income side. You've got to get paid, and you must collect "receivables" on time and in full. As I was jolted into understanding a few years back, receivables are distressingly easy to lose track of. I had just completed the final item on a client's punch list when she said, "I'll get you that $2,500 right away." Before I could restrain myself, I blurted out, "what $2,500?" With that episode, I decided to create a receivables journal.

You can purchase a receivables journal, complete with pegboard and blank invoices, that works on a one-write principle similar to the disbursements journal with checks. For my work, however, I have found it more cost effective to create a receivables binder from a three-ring binder and standard stationery-store spreadsheets. For each project, I set up a page in the receivables journal to record payment, and calculate the balance as work proceeds. At the end of the project, the balance must be 0 or I have collected too little or too much. Once, on a quarter-of-a-million-dollar project, I overcollected $10,000 near the end of the job. I stared down temptation (Spring in Paris? Winter in Rio? New basketball court in the backyard?) and sent the client a refund check, counseling myself that honesty pays and that she would refer good projects my way. She has.

A receivables journal only enables me to determine that I have collected the proper payments. To make sure I get paid in the first place, I collect from clients personally at the project site. I do not bill my clients by mail. The payments due small-volume builders are too big—often tens of thousands of dollars—to entrust to any delivery service. Personal collection also lets you know at the earliest possible moment if you will have difficulty getting your money. It prevents the occasional shady client from claiming an overdue payment is "in the mail." In addition, personal collection allows you to get your money into your bank account quickly, where it earns interest and beefs up your working capital. If handled properly, personal collection does not amount to badgering your clients. You let them know in advance when a payment is due. Then you use the collection as an opportunity to review the project's progress and to show the clients how they are getting their money's worth.

CONSOLIDATION AND SOPHISTICATION

To manage your company in accordance with the crucial principle of minimizing overhead, you must control your use of time as successfully as your consumption of equipment and material. Every one of your procedures will tend to become lengthened with unnecessary steps, using more in energy than they give back in efficiency, unless you periodically tighten them. You will see the need for clear and compact procedures when you hire a bookkeeper, a step you will likely find cost effective after a year or so of doing the books yourself. A bookkeeper who steps into a loose system cannot work efficiently or accurately.

Hire a bookkeeper as carefully as you hire any other professional. Ask other builders and your accountant for recommendations. The bookkeeper you choose should have neat handwriting, several years of experience (including experience with pegboard systems if you use one) and a commitment to the work. Hire good, not cheap. Erroneous books can lead to management mistakes far more costly than the extra pay a good bookkeeper commands. To work efficiently with your bookkeeper, consolidate your paperwork procedures into a sequence such as this:

1. Designate a weekday morning to collect time cards and to phone your employees' hours to your payroll service.
2. Designate the following morning (when paychecks with reports and returns will be delivered to your office by 24-hour courier) for your paperwork. Sort out the contents of your in-box and handle accounting tasks first, while you are freshest.

3. Review for accuracy all invoices from subs, suppliers and services. (You may find a surprising number of mistakes. One medium-volume builder claims to cover the entire salary of a full-time bookkeeper with the savings from the overcharges found in invoices.) Also review your payroll checks, reports and returns.

4. Sign the payroll checks for delivery to your employees at the job site.

5. Sign the checks from clients for deposit.

6. Organize deposits, bills, reports, returns and other necessary material, such as bank statements, for your bookkeeper.

7. Do your job costing. Bring your job-cost cards up to date from the invoices and time cards. Make out unit-cost sheets for completed phases of projects.

Gerstel's complete bookkeeping system: a one-write income and disbursements journal with bank of checks, envelopes and field checkbook; a file of completed disbursement pages; a binder with cumulative records by quarter from payroll service; colored envelopes to store invoices (with check duplicates stapled to the back) and time cards for each project; a receivables journal to record collections on each project; two additional files, one to hold records and duplicate checks for all tax and insurance payments and another to store records of all other non-job cost payments.

After the bookkeeper has completed his or her work, go over it carefully. Make certain check amounts correspond with invoices. Review entries in the disbursements journal to ensure they have been made according to your chart of accounts. (Even experienced bookkeepers occasionally make mistakes. Minor ones may hardly matter, but the incorrect entry of a large check can throw off your records and, therefore, your management decisions significantly.) Sign checks and returns, and mail them.

One final task: Make sure your bookkeeper is not dipping into the till. Determine that all checks in the numbered sequence have either been used to pay legitimate bills or marked void and stored in a designated envelope. If any are missing, find out why. If you have any reason to question your bookkeeper's honesty, watch your bank statement for signs of possible theft, and talk with your accountant about appropriate methods for safeguarding against embezzlement.

With your bookkeeping routine complete, you are ready to analyze the data it has produced. From your job-cost cards, you can discover the need for changes at the project site. For example, is the trim only a quarter complete while the budget is half used up? Perhaps your crew is working to stain-grade instead of paint-grade standards. Likewise, examining the figures in your disbursements journal can help you recognize a need for innovation or correction in the overall management of your company. A first question to ask: Do any of the totals for job costs, labor, overhead or other columns seem high? Here are some examples:

- Capital investment. Has it surged to 20% of your gross? Is all that new equipment earning its keep on your projects? Or is it just consumerism disguised as business investment?
- Crew supplies. Has it risen to over 10% of gross wages? Maybe your leads have been tacking on too many nifty small tools when they place their orders at the building-supply company.
- Insurance. Are your premium installments running way ahead of what's necessary for your level of payroll? Or, on the other hand, are they running behind, so that you will face a disruptive balance due at the end of the year?

Each column in the journal deserves similar consideration. Likewise, subtotals of the column groups—job costs, labor costs, fixed overhead costs—raise important questions. What, for example, do you make of your fixed overhead? Is the total a lower percentage of your gross receipts than you expected? Can you afford to tighten your bids and become more competitive? Or are you working too cheaply, not charging enough to cover overhead and provide a reasonable profit?

After analyzing your costs, turn your attention to income. To check your earnings to date, make a two-step calculation. First, figure earnings showing in the journal:

Deposits + non-deductible checks –
total of all other checks = journal earnings

Second, adjust for receivables due but not recorded, and bills (payables) owed but not paid, to bring your figure up to date:

Journal earnings + receivables due –
payables due = actual earnings

Oh-oh. Bad news. You see that you've made less than your lead carpenter (a frequent experience for small-volume builders). It's time for a management decision. Maybe you conclude that you have been too heavily involved in glamour jobs, with those enticing but never-ending and labor-consuming details. You decide to pull in a few lucrative, straightforward structural-repair jobs to leaven your loaf.

With a calculation similar to that used for earnings, you can project your cash position:

Cash balance + receivables –
payables = cash position

In other words, you know what you have in the bank, what you are owed and what you owe. With the requisite addition and subtraction, and assuming your clients are good for the money, you can project your cash position. Perhaps you realize that upcoming bills will cause you to overdraw your account. Must you line up a short-term bank loan? Or can you devise a strategy to avoid that overhead cost? With an eye to the longer run, can you install a program to build up working capital and avoid future cash crises?

If you keep up your disbursements journal, do job costing, use a payroll service, file accurate tax returns and thoughtfully analyze your numbers, you have come a very great distance in the financial management of your company. You should know, however, that you can go much farther. To begin with, you can move from cash-basis accounting to the accrual system. Then costs and income are recorded when incurred or earned, not when paid or received as with the cash system. One accounting expert argues cleverly that "reporting on a cash basis in a charge society is not realistic." His argument might be important for you if you build for developers or do other work where payment can lag far behind performance, so that you in turn must pay your subs and suppliers long after receiving their bills. In that case, unless you set up accrual accounting, you might have little idea of your cash or earnings position. Also, at some point in the growth of your business, you might need to move to accrual accounting to satisfy the IRS; they might consider that cash accounting gives an inaccurate picture of your income.

On the income side of your accounting, if you do a substantial volume of time-and-materials (T&M) work, you will need to institute controls to make sure you are billing for all your costs. You can bill strictly off invoices and time cards. (I take the risk, since I contract for very little T&M work.) But if you lose a document, you will likely not notice and be out the money. A company that does a significant amount of T&M work therefore needs a more sophisticated billing system with built-in proofs. Such a system, however, is beyond the scope of this book. If you need it, seek help from an accountant and/or do further research.

On the cost side of your accounting, equipment records are a tool you will find essential if your work requires costly machinery, such as a backhoe. With equipment records, you track purchase payments, storage, maintenance and other costs of ownership against hours of operation and get a view of the cost effectiveness of your investment. You might learn, for example, that the hourly cost of operating your own backhoe is twice that of renting as necessary, and that as much as you love it, you should sell that hoe.

If your work requires construction loans or performance bonds, you will need financial statements. Occasionally, in my experience, even banks making a loan for a residential remodel ask for a financial statement to ensure the client is engaging a solvent contractor. A financial statement summarizes the amount you own (assets), the amount you owe (liabilities) and the resulting balance, your net worth. Some agencies will accept a financial statement you have drawn up yourself. Others will require that the statement come from a certified public accountant.

Beyond accounting tools such as accrual systems, equipment records and financial statements lies a more sophisticated form of accounting known as double-entry bookkeeping. Among its advantages are that it minimizes the possibility of mistakes and embezzlement. But as the book *Small Time Operator* (see Resources, p. 223) points out, it also "transforms business bookkeeping from a part-time...to a full-time occupation."

Unless you want to be a full-time bookkeeper or feel the rapid growth of your company requires you to hire one, as you add new accounting features remember to keep your system appropriately simple. Ask whether you're adding a feature merely because some accountant or bookkeeping manual says you're supposed to or because the new data will actually help you make better decisions. Will it significantly reduce your chances of error? Will it improve your position with the IRS? Will it enable you to control your costs and monitor your income and cash position better? Will it allow you to guide your projects and support your crew better? Will it allow you to serve clients better? Or is it only going to bog you down in marginally useful numbers when you could better spend your time winning and running construction projects?

PART 4

GETTING THE RIGHT JOBS

Promotion
Prospects
Qualifying Projects
Qualifying Clients
Working with Architects
Competitive Bidding
Price Planning (Negotiated Bidding)

PROMOTION

When you're ready to go into business for yourself, you may find that you have already built a bridge to full-time self-employment. As you were learning your trade, you may have been taking side jobs for friends and acquaintances. Your first customers may have recommended you to others, and they to still others. Suddenly, with surprising speed, you're pulling in projects through a widening network of references and are on your way to the traditional small-volume contracting business.

Even if this delightful scenario is yours, however, at various times you may find yourself having to hustle after work. Some possibilities include:

• Real-estate offices. Realtors have the reputation of seeking cheap, (not good) construction. Give them good work anyway, and your performance will likely bring in customers who pay fairly.

• Remodeling companies. Some storefront remodelers run their own crews. Others are really retail operations. They sell cabinets and finish materials, but refer customers to contractors for installation and any necessary remodeling.

• Department stores. Outfits like Sears and some big hardware stores offer a full line of repair, renovation and remodeling services with the construction subcontracted to small-volume builders.

• Public bids. Public institutions, including city-housing departments, the post office and colleges, regularly put projects out to bid, or for their smaller jobs, invite a trusted contractor to do the work for a negotiated price.

- Corporations. Banks, hospitals and other large companies often need reliable contractors.
- Insurance adjusters. Some insurance work demands specialized skills. Fire-repair work, for example, requires that you know how to seal damaged framing to suppress the acrid odor of smoke. If you go after insurance work, take care not to overshoot your competence, which could land you in a lawsuit.
- Adult schools or owner-builder centers. Teaching classes at these institutions can introduce you to prospective customers.
- Local builders' associations. You can meet well-established contractors at meetings. In time they may pass on to you the smaller jobs they are too busy to handle.
- Architects. Working with architects is so complex that it is covered separately on pp. 93-99.

Perhaps the best way to drum up work is by canvassing. In some locales, going door-to-door will reportedly bring swift attention from the sheriff. In others, if you have the knack, you can bring in substantial business. Wherever you work and live, you will have continual impromptu opportunities to promote your business among people you chance to meet. I relied on this technique heavily in the early days of my business and soon learned that almost everyone has a construction project that needs to be done, and that small jobs often lead to much bigger projects. Whenever any passersby showed the slightest interest in my work, I lured them onto the job site for a look around, and made sure that when they left they had my business card. Whenever I met anyone at a party, in a coffee shop or on the street, I mentioned that I was a builder. (What do you do? Oh, you design computers, that's interesting. I'm a builder myself. And how long have you been designing them? Eight years! Sure, I remodel bathrooms.) In later, more prosperous years, I would brag that my company's success was based solely on word-of-mouth. "Yes, your mouth," my wife liked to remind me.

Once you actually begin a project, your best promotion opportunity comes from doing a good job. Of course you must build plumb, square, straight and tight with sound material for a fair price and on schedule. But for the sake of promotion, less obvious practices count as much. It's especially important, for example, to keep a clean, orderly job site. With debris- and litter-strewn sites the norm, a well-organized project distinguishes you as a class act. I regularly get calls from prospective customers who have been impressed by one of our sites—they take its appearance as a sign that we do careful work, too.

Ranking in marketing power with the well-kept site is courtesy. Courtesy, as I've come to understand its meaning in construction work, includes not only politeness, but also encompasses communicating fully with your clients. It means letting clients regularly know what is happening and why. It means answering their questions patiently and answering their phone calls immediately. It includes taking the trouble to clear up misunderstandings — and even productive relationships have an occasional glitch — with a frank and friendly chat. People are loathe to give repeat business to builders who are discourteous or not thoughtful enough to keep them informed. One survey reports that among people who have decided against working with a contractor they have employed before, fully 70% gave as their reason the contractor's discourtesy.

Too often construction sites resemble not the workplaces of well-organized professionals, but the public dump. Incredibly, builders often place signs advertising their businesses in the midst of these eyesores. Do their signs actually draw clients? Or do passersby instead make a mental note never to call the people who created that ugliness?

Larry Hayden, owner of the 65-year-old Federal Building Company in Oakland, California, has this to say about promotion: "You're much better off spending your time with an existing client than you are going out and looking for ten new ones. They're your best salespeople." But references from satisfied customers — that precious "word-of-mouth" business, are not chance events. If you do your job well, they will come your way. When people successfully complete a construction project, no mean feat, they want to tell their friends. They are proud that they managed to get it done right. They also appreciate your diligence and are often eager to give back something more than mere payment. A vigorous referral to a friend is a way of saying "thank you."

Once a job is done, a customer's enthusiasm may slowly dim. You may wish to fan the coals. All you need do is to keep doing your job. For example, call your customers back three and six months after completing their jobs to make sure all the work is in good order. After a year, call again. Ask for an evaluation of the project and your company's performance. If anything is wrong, go back and fix it immediately.

You may hesitate to call past clients, for fear that you will spark a callback. At the start-up stage of my career, when all I could think about was getting on with my current projects, I hated them, too. But over time, I've learned that callbacks are not a burden, but an opportunity for the most cost-effective promotion possible. Nothing revives a customer's appreciation (and the impulse to pass on your name) like your lightning-quick response to a problem.

The best reputation will do you little good if prospective customers cannot find you. Invest in a few inexpensive tools to make yourself readily accessible. A business card is essential. Giving your name and number is often the first exchange with a prospective client—make sure you don't have to scratch around for a pencil stub and paper scrap to do so. Keep your card with you at all times (your crews should also have a supply on hand). All documentation, such as contracts, should include your name and phone number. Your name, properly spelled, must be listed in the phone book. And don't forget to turn on your answering machine.

Job-site signs may have value in maintaining your accessibility, though I question whether they are more of an ego tool than a productive business investment. With so many contractors' signs dotting a town, the public grows blind to all but the most striking. Builders I have asked tell me their signs rarely if ever directly bring them work. I do not use job-site signs (nor the even more suspect truck sign). Given their initial purchase, maintenance, storage and handling costs, signs are a significant overhead item that I don't believe to be a sound expenditure. If you decide to use signs, take the trouble to produce a design that is noteworthy and legible. I often see signs with oh-so-subtle colors and delicate lettering that can't be read from distances greater than 15 in.

With a commitment to good work at all levels, attention to callbacks, a distinctive business card, an answering machine and perhaps an eye-catching job sign, you have likely gone as far as you should with promotion. Advertising in the usual sense of the word—reaching out via the media to an anonymous public who has no word-of-mouth or direct knowledge of your work—is of dubious value. I'm not alone in my skepticism. Deva Rajan, a builder with 30 years of experience who has been ranked among the best in the country, says of advertising: "Stop it. Service your product and you don't need it." Steve Nicholls, owner of a cabinet company with 25 employees and a reputation for excellent work, warns that "Advertising can deteriorate your client base. It leads to cut-rate customers who lead to other people looking for a cut-rate deal."

The most convincing case against advertising comes from Salli Raspberry and Michael Phillips in their first-rate book, *Marketing Without Advertising* (see Resources, p. 223). The typical American, Raspberry and Phillips point out, is bombarded with 1,600 ads a day. The sheer volume makes mockery of the notion—one I often hear small-volume builders use to justify their investment in ads—that there is somehow value in "keeping your name before the public." Perhaps a national brewing company can invest enough in ads to make a significant dent in public consciousness. But what are the chances of your voice being heard amidst the general din of hype?

Business cards can use photos or drawings or can be composed completely of type. Try to make yours distinctive, without going to great expense.

Even if you do draw prospects with your ads, they may not do you much good. According to research, says Raspberry and Phillips, customers lured by ads are not loyal. In fact, because much advertising is deceptive, companies who use it are suspected of offering shoddy goods and services.

Phone-directory display ads (that is, the Yellow Pages) deserve special comment. Although builders often assume these are a must, one builder I know calls his directory ad "the worst mistake I ever made. It got me involved for endless hours with tire kickers who were really just looking for free advice." Another builder reports that during a single year, he met with 320 people who called in response to his ad, got one job for his efforts and did not make enough on it to cover even the cost of the ad. Still another builder keeps himself listed (just a name and number, as opposed to a display ad) in the commercial directory. But he declines to talk with people who call him cold, politely explaining that the listing is simply for the convenience of his regular customers. He has found that he gets too little profitable work from unknown callers to make talking with them worthwhile.

Still, a very good business manual written by two painting contractors *(Paint Contractor's Manual,* see Resources, p. 223) calls a phone-directory display ad essential, and it might have value for other types of contractors as well. If you do consider investing in an ad, first try to calculate the potential results. Talk to local builders who use a phone-directory ad. If you don't know them personally, call them and ask if their ads bring enough business to justify the cost (including the cost of responding to all the tire kickers). Don't be swayed by vague responses, like "Well, the ad keeps my name out there." Ask for specific results.

In addition to phone-directory ads, you may wish to evaluate the following types of advertising:

- Classified ads. These can't hurt as much as directory ads, because they cost far less and are not seen by as many people. If you are new in town, classifieds may help you get started. Although they are likely to attract cheap customers, you can work up from them.
- Flyers. Many builders distribute flyers to neighbors when they begin a new project. Flyers are inexpensive and can bring prospects. But they can also attract the neighborhood sidewalk superintendents who will distract you and your crew.
- Display ads in the media. Some builders report cost-effective results from display ads placed regularly—not just once or twice—in controlled-circulation media, such as weeklies that go only to homes in selected neighborhoods.
- Home shows. Including your time, the cost of creating a booth and fees, a home show can be a pricey way to advertise your busi-

ness. But if you are good at talking about your company to strangers, shows may generate promising leads.

- Postcard mailers. Of all the ads, this type may be the most worthwhile. Again, do your research and monitor results closely. But with a postcard or one-page mailer, you need not solicit an anonymous public. Your mailer can go to past clients and to others who already know about your good work. Rather than hyping the masses, you are rejuvenating your word-of-mouth reference network—a truly vital source for potential new projects.

 # PROSPECTS

First impressions are tough to reverse, so when prospective clients call, you want that first conversation to be crisp. With a checklist such as the one on p. 82, you demonstrate that you are well-organized. At the same time, you show interest in the project and begin to determine whether it is a good one for you.

As you get acquainted with the client during your initial phone call, try to avoid giving an off-the-cuff cost estimate. It took me years to learn my lesson, and now I never give one. If your figure is low, the client may rely on it and waste money on a design and working drawings for a project that will never be built. I have seen this happen again and again. If it is high (meaning realistic for good work), you may be scratched off the list of potential contractors. I have had clients scream at me when I told them roughly what a project would cost, and they did not apologize later when my projection turned out to be correct. Quite understandably, clients often push for that "ballpark estimate." But if you give it, you're stuck with it. Instead, you must explain that meaningful construction cost estimates must be carefully put together. Any quickie figure is likely to do damage, maybe to you, maybe to the client.

In addition to holding the line against requests for cost estimates, during the initial phone conversation it's best to avoid negative responses. I've made the mistake of becoming negative too quickly and losing opportunities. At other times I have kept my counsel and ended up doing jobs that originally appeared unsuitable. Clients who said they wanted to start in six weeks have waited six months for us. Others who had named as their absolute top figure a budget half of what seemed minimally necessary have come up with the money when they saw what it would take to do the job right.

Unless the project is clearly too small, I usually make an appointment to meet with the clients in person. For that meeting I have two rules, both learned from a remodeling salesman with three decades of experience. First, if the clients are a couple or a partnership, both

Both postcards shown here distinguish themselves from the average run of junk mail that piles up in the mailbox each day. Strictly Custom's card (at top) reminds customers of their devotion to technical excellence. The card from Mueller Nicholls (above) underscores the unfailing good humor of the proprietors.

Checklist for Phone Interview with Prospective Clients

Date
Names (husband and
  wife/all partners)
Phone
Address
Referred by
Description of project
Current status of design
Design-build services desired
Architect or designer's name
  and phone
Current status of permits
Budget
Acceptable starting date
Acceptable completion date
Bidding process desired:
  Competitive
  Price planning (negotiated)
Number of bids sought
Date bids due

husband and wife or all partners must be present. You can spend a morning working with a prospect, then be dropped from consideration because a point you've made is incorrectly transmitted to the absent individual. Worse, you can move toward bidding (and, still worse, winning) a project from which you would have recoiled had you met the absent person.

Second, I require that the meeting take place during the fat part of the work week—8 a.m. to 6 p.m., Monday through Friday. I decline weekend appointments except in unusual cases; I'm looking for clients who respect me, and a good indication is whether they respect my time off. On weekdays I do not want our meeting to be shoved into the leftover time after work and dinner. To customers who hesitate to take time off from their jobs, I point out that if they select my company to build for them, we may be working together for months. We need to meet at our best, not when we are tired after a long day. Clients usually understand. After all, they take time off to go to the dentist, and a construction project is far more expensive and potentially far more traumatic than a multiple root canal.

When you arrange to get together with clients, ensure that they have budgeted ample time for your meeting. (I usually ask for one and a half to two hours, depending on the size of the project.) Don't get on an assembly line with a row of other contractors. Arrive at the meeting a couple of minutes early; that's much better than calling ahead to apologize for running late. You will immediately distinguish yourself from the contractors who are not punctual.

Proper dress for a meeting with clients is a matter of amusingly heated debate. (The subject seems to touch that collective sore point: Are builders professionals or proletarians?) Personally, I like freshly laundered jeans and plaid shirts—my image of the "real" builder. But contractors I respect insist on a coat and tie, saying such "power dress" puts them on equal footing with white-collar professionals and gives them credibility with all clients. Down at the other end of the spectrum is the successful commercial remodeler who works largely in San Francisco's swank financial district. He wears his dusty, stained work clothes to meetings with clients, figuring that they see him and think, "Here's a guy who works for his dollar. I'm getting my money's worth." I suppose the lesson here is, if you feel comfortable and confident in whatever you are wearing, your clients will feel confident in you.

For the first meeting, as for the first phone call, it is important that you have a clear agenda. My top item is listening. I want to deepen my knowledge of the clients and the project. Exactly what do the clients propose to build? Are the plans realistic? What is their attitude toward me? Productive listening requires not passive acceptance of information, but responsive attention—smiles, nods and enthusiastic encouragement. "Try to imagine," suggests cabinetmaker Steve Nicholls, "the position of the customers, knowing nothing, confronted by an overwhelmingly knowledgeable builder." Clients can easily feel amateurish and silly as they explain their ideas. Active listening can put them at ease and, moreover, is a good sales technique. Once I asked a customer why she and her husband had chosen my company over the respected contractor recommended by their architect. She said it was because I had listened enthusiastically as they described their ideas and hopes, while the other contractor had pushed them along, impatient to get to his own presentation.

When the clients seem to be winding down, I suggest that I take a turn to explain how my company operates. I begin by explaining to the clients why they should be looking for a builder with a dual capacity—good management supporting good craftsmanship. To show my company's capacities, I take the clients through a portfolio of our work, explaining our procedures, skills and resources as I go.

The first part of the portfolio consists of a single volume that demonstrates management and organizational skill. It displays in logical order the various documents—contracts, flow charts, "Do" lists, change-order forms, etc. (all discussed in coming chapters)—that ensure full communication and tight management throughout our projects. Interspersed with the documents are photos of actual construction. Several examples of each stage of work, from foundations and frames through finishes, are exhibited. I point out a few of the elements that make for a durable final product. "Finish work," I like to emphasize as I show photos of accurately installed rebar, void-free stem walls and careful flashing, "begins at the foundation." A project can look good when it is just completed, but without sound rough work underneath, the finish will begin to deteriorate in a few years. I also point out our efforts to protect the homes and gardens of remodeling clients and to keep our sites neat, clean and safe. I describe our measures to minimize waste and to salvage or recycle virtually all scrap material. I enjoy recycling, but it also happens that people have become widely concerned about ecological problems. As a result, I have won jobs because my clients warmed to my ideas for minimizing the environmental toll their project would take while my competitors ignored the topic.

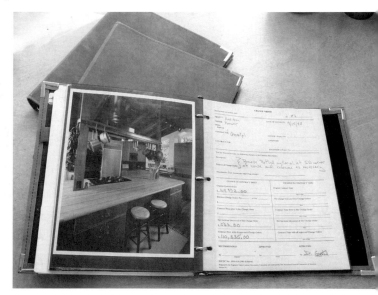

The first volume of Gerstel's three-volume portfolio intersperses samples of project documentation such as contracts, change orders and flow charts with photos of construction work in the sequence it is typically performed. The other two volumes display photos of a variety of completed projects.

My management-sequence volume may be somewhat unusual. The rest of the portfolio is more traditional—two volumes of before, during and after photos of projects. I have found it best to hand the volumes to the clients and let them look at their own pace, while I point out projects similar to theirs and explain how quality was achieved. I introduce the crew and subcontractors who appear in the photos, telling a little about the skills and personality of each.

At the close of a presentation, you want to leave your prospective clients with some promotional material. Good possibilities include a map showing the locations of your completed projects and a page of brief crew and subcontractor biographies.

But the most valuable item you can leave is your list of references. If you are new in business, create a list of former employers and other professionals who can attest to your skill and character. Once you begin completing projects for your own clients, you may wish to include them on your growing reference list. My list, as I take care to explain to new prospects, includes virtually every client we have built for, not only those who will assuredly give the company a rave. Such a complete list inspires customers' trust, even if they may hit an occasional off note. Indeed, the sour note may make for better music. One of the cardinal rules for making references effective is to mention a negative or two along with the positives to make the person real.

Often at the close of the first meeting, clients again raise the question of cost. Again I explain that arriving at a meaningful estimate requires careful work, and that by throwing out a figure that could easily be wrong, I may do both of us a disservice. I then turn the question around, asking for the clients' budget, so that I can judge whether they are being at least remotely realistic. Unless the project is extraordinarily complex or unusual, I can generally offer one of three responses. First, the budget is generous (rare). Second, it is adequate for the proposed project, though perhaps not with every item on the wish list (usual). Third, it is not enough to do a good job using a legitimate contractor, and if the clients try to do otherwise, they take serious financial and legal risks (occasional).

I have never had a client refuse to reveal the budget, though some have hesitated for fear of encouraging an inflated price. To reassure them, I show a sample of the detailed estimate breakdown (p. 118), which helps customers see my prices are not padded—that material, labor and subcontractors are charged at cost, and that the markup is reasonable. I must know the budget, I explain. Otherwise I cannot go on to the next step, determining whether it's worthwhile to produce an estimate and bid for the project.

QUALIFYING PROJECTS

When a prospective job looms on the horizon, your natural inclination is to chase it down the track. Settling back into the starting blocks and letting one go by is hard, especially at the start of your career or during a recession, when you are not confident that work will keep coming in. However, cannily discriminating between projects and selecting the right jobs to pursue is a skill you must hone. Otherwise, you will squander time and money pricing projects that will never be built, that you have no significant chance of winning or—worst of all—that will overwhelm you if you do get them.

Before you spring into pursuit of a project, you want to see several starter flags come down. First, the budget must be reasonable. If you didn't learn the budget at your initial meeting with the clients, go back and get it. Also find out where the money is coming from and where it will be during construction. Builders often hesitate to inquire about the clients' money, as if they were somehow prying into personal affairs. On the contrary, when you ask about budget and funding, you are inquiring about a subject intimately connected with your own well-being. "Many contractors," Associated General Contractors points out in their *Basic Bond Book* (see Resources, p. 223), "have gone broke because they didn't ask where the money was coming from to fund private jobs."

Even if your survival is not at issue, you risk wasting time when you ignore budget constraints. For example, some years ago I was asked to bid the construction of a large, elaborate home requiring a massive foundation to hold it on a steep lot. I inquired about the budget and was told it was $250,000, which worked out to a square-foot price just a little over the going rate for "slapstick construction," as one of my apprentices nicely describes tract housing. I declined to produce a bid, figuring I would invest a great deal of work with no chance of return. Four other builders did submit bids. The two good builders came in at about $500,000, the two mediocre ones around $400,000. The owner turned all four down, went shopping again, and finally ended up acting as his own general contractor. I asked the builder I regarded as the most capable of the four why he had gone to the expense of submitting a bid for such an underfunded project. "I didn't ask about the budget," he answered. "I guess now that you mention it, I wish I had."

Once budget and funding have passed scrutiny, move on to the next qualifier—schedule. You need adequate time both to prepare your estimate and to build the project if you win it. There's no point in spending days on a bid only to have it rejected because you

Checklist for Qualifying Projects for Bid

Don't jump to bid on everything. If you do, you will waste time and money pricing projects that will never be built. A checklist can help help keep your decisions objective.

Bid due date
Required construction schedule
Budget
Quality of design and working
  drawings
Your ability to manage
Crew leader availability
Crew availability
Subcontractor availability
Supplier availability
Equipment availability
Distance from office
Commute for crew
Availability of utilities
Portion of work to be
  subcontracted

missed the deadline. Nor do you want to rush to meet a deadline; to catch mistakes after completing a bid, you need to set it aside for a few days and then take a final fresh look at the numbers. Likewise, you do not want to obligate yourself to a construction schedule you cannot comfortably meet. Either hurrying a job to completion or running behind can cost you money and reputation. In qualifying a project for schedule, do note, however, that clients are often naively optimistic about a starting date and construction time. If their requirements at first seem unacceptable, stay in touch anyway. You may find that reality has set in and modified the schedule to something you can accommodate.

Unless you will be providing design services yourself or have built with the designer before, you should evaluate any drawings as part of qualifying a project. They must be clear, legible, accurate and workable. Too often, drawings are so crammed and cluttered you miss information. If they are not legible, you will misread them during estimating, as I did recently when I assumed the inkblots on a shear wall schedule called for the usual nailing, not three times as much. When dimensioning is not accurate and consistent (for example, an architect has figured 2 + 3 = 7 at some crucial juncture and the error dominoes through the layout), you can see costs soar as your crew sorts out the adjustments during construction.

Just as the numbers must be correct, so must the working drawings actually work. A builder I know did not notice until well into framing a particular roof that the design resulted in 6-ft. high bedroom doorways. He had to take his work apart and start over. Finally, working drawings must be complete. All the necessary plan views, exterior and interior elevations, structural sections and finish details must be present. When I qualify projects, I particularly look for complete door and window schedules. If these are absent, my crew and I will have to invest a lot of time putting them together. Moreover, designers who provide the schedules and get them right usually have their other bases covered as well. If the drawings seem of dubious quality, you will want to invest extra care in qualifying the architect or designer and engineer themselves. But that subject will be covered in more detail later (p. 93-99).

Even if a project has been designed and drawn by competent people, is reasonably funded and incorporates realistic bidding and construction schedules, it still may not be right for you. It must also fall within or be reasonably close to your company's niche. Factors you will want to consider include:

- Your personal ability to manage the project. You may be an ace residential remodeler. But that does not mean you can successfully bid and build a new restaurant in an existing commercial complex or a large custom home.

Clear Plans are Essential

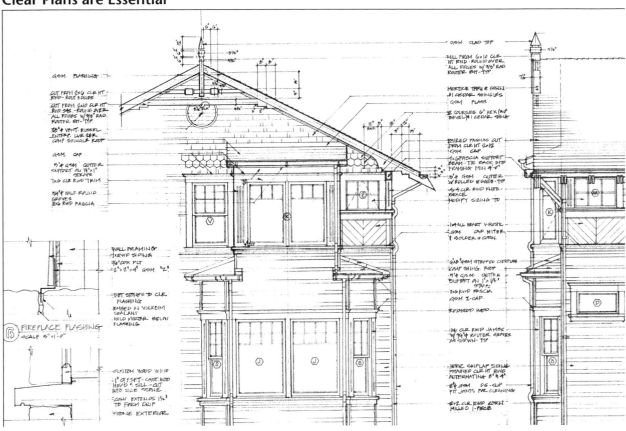

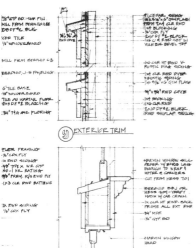

- Availability of a capable lead. Ideally, you already have the right person employed in your company and in sync with your crew, subs and suppliers.
- Availability of crew. Hiring new people takes time. Fitting them into your operation slows you and your regular people. If you must expand from, say, three to eight for a project, your productivity per person may decrease.
- Availability of trusted subcontractors. You can introduce a couple of new subs on a single project. But if you are not an established, and therefore priority, customer for the majority, your project can experience serious bottlenecks when several of the new people simultaneously put you on their back burners.
- Familiarity with suppliers. Learning a supplier's methods and finding the right staff to deal with take time. As with crew and subs, you want to try only a few new suppliers on a given project. For any technical enterprise, including construction, it makes sense not to change too many systems or components at once; make changes one sure step at a time. In other words, prefer evolution to revolution.

David Ludwig's plans are known for their legibility and accuracy. Unclear plans can foster ruinous error during estimating and construction.

- Distance. If a project is too far from your office, visiting the site, especially on short notice to solve an urgent problem, becomes a burden. If it is too far from your workers' homes, their energies go into driving instead of building. If it is out of your subs' and suppliers' usual range, they may be slow servicing the project.
- Availability of water, power and phone. If you are accustomed to working in an urban community, you may find yourself struggling if you take a job in an outlying area without the utilities you take for granted.
- Portion of project to be subcontracted. By subbing out work, you share risk. If your crew handles all the work, you take all the risk.

As you gain experience, you will be able to qualify or disqualify most potential projects with a quick run through a mental checklist. But if you are in doubt, if the project falls in a grey area, you can sharpen your perspective by sketching a bar chart. Recognize, however, that checklists and charts, while lending objectivity to your decisions, cannot eliminate subjectivity. If a project bores you, you are likely to deflate its rating; for the exciting job, you may unconsciously inflate the score.

Start-up builders especially must resist swooning at the siren song of the "glamour job"—one that is substantially bigger, fancier and more prestigious than any previous project. "Often I have been so eager to take on a project that I would do less qualifying of it than I

Qualifying a Project Using Graphs

Evaluation — Smith Addition, Oakland

With a simple series of bar graphs, you can determine if a project is right for you.

	Score	Comments
Ability to manage	9.5	Right up our alley
Crew leader	9.5	Fred skilled at all phases
Crew	8	Will need to find another good framer
Subs	8	All regular subs available except plumber (vacation)
Suppliers	6.5	Too far for regular concrete supplier
Distance/Commute	6	At the limit of our comfortable range
Overall	7.9	Go for it!

Does not qualify (0-4) | Marginal (5-7) | Qualifies (8-10)

should have," I was told by a builder who had just spent half a year working at minimum wage because he had badly underbid an appealing project. "I have learned to be realistic, not to become so enamored of a project that I kid myself about what it will cost me to do it."

Just as you should not subjectively overqualify the glamour jobs, do not shun the small, plain ones. They will keep you and your crew employed and cover your overhead while you wait for something better to come along. If you are bidding tight to stay busy, better that you make those bids for small projects with limited potential for losses. And consider this: Small projects can contribute mightily to the long-term growth of your business.

If you do one big project in six months, it will add only one client to your word-of-mouth network. But with six small jobs you can gain six satisfied clients, each of whom can be the beginning of a whole new branch of the network. Recently my company completed a ten-month project that earned us a solid profit and received an award for quality of construction and design. We were the only contractors considered for the project; no others had been asked to bid. When I traced the project's lineage, I found myself working down a chain of projects until finally I arrived at a two-week foundation-repair job I had taken on six years earlier to keep a crew busy.

↑
Levinson house

↑
Granal second-story addition

↑
Johnson kitchen

↑
Alvarez foundation

↑
Jones deck

QUALIFYING CLIENTS

You can't size up people as systematically as projects. Checklists and graphs won't do you much good. You must take intuitive soundings for broad conditions like mutual trust and affinity, which are the prerequisites for complete and clear communication during a project. When builder John Larson first visits a client, the response he looks for is respect. "I can't be treated like hired help," he says. "They are not just going to direct me to put some wood together." He wants clients who respect not only his trade skills, but also his skill at putting the whole package together. Some clients, even while appreciating good construction, will not understand the need for management. You will know them by their question, "Do you do the work on your jobs?" They want to know if you hammer and saw. If you say no, they wonder why they should pay for your role in the project. Since they think of construction as labor and materials only, they do not appreciate what it takes in effort and overhead to organize a building company, to create and sustain all the many relation-

A small job can often generate the first in a series of references that leads to that coveted big project.

ships with crew, subs and suppliers, to design and mesh all the systems, and to keep all those people and systems successfully in motion for their project.

There are a couple of answers you can give such people. The one I like best is that, yes, you may do some hammering and sawing, but whether you do is not really important. For if you do play a hands-on role at the job site, your hands will still be only one pair among many putting lumber, steel, wire and pipe into place. They will be the only pair, however, guiding the machine that executes the project. This is your crucial role, to make sure the right people and the necessary materials are at the project at the right time with the necessary support.

If that answer doesn't work, you can encourage the clients to act as their own contractor, suggesting that if they do everything right, they may save part of your markup and make the minimum wage for their efforts. With a few missteps, they will end up spending more for an inferior product and spend many anxious days and nights wondering if the whole project is going to unravel. As they consider the prospects, the clients may get the point: Construction management is difficult work. It is worth paying some fraction of a project's total cost to make sure it is built properly.

Even customers who respect both your trade and management expertise may want to "co-build" with you. Reasonably enough, they want to save money, and they imagine that they can by directly supplying products to the job ("I've got this great connection for cabinets"), or by handling a subtrade ("My brother is a plumber in Laramie, Wyoming, and he'd be glad to fly out for a weekend and put in the pipes"), or even—heaven forbid—by working alongside your crew. They do not understand that by substituting themselves for a professional sub, supplier or carpenter, they are forcing you to deal with amateurs. They do not appreciate that while they may save themselves money, it will be coming out of your pocket.

Some builders flat out will not co-build. In most trades, in fact, customer participation is discouraged. (I once came upon a sign in an automotive shop that posted hourly labor rates as follows: If you leave: $40; if you watch: $50; if you help: $60.) For my projects, I discourage owners from participation, pointing out that they will likely earn less than minimum wage for their efforts, produce mediocre results and endure much frustration. Nevertheless, some insist. They feel, occasionally rightly, that from working with Grandpa as a kid or from years of fixing up around the house they have acquired serious skills. Sometimes, they just want the satisfaction of helping to build their own place. For those people, I have settled on an approach that allows owner participation, but protects me from losses. I require that owners do their work either before or after my crew arrives at the job. For example, owners can gut their kitchen before my crew and subs move in to build the new one. They are required under the contract to perform their tasks so as to mesh with our schedule. When we arrive, if any of their tasks are incomplete, we finish them at extra charge. Similarly, the owners can get involved at a later point in the construction sequence. But at that point we leave, and they must handle all further work. If they want to provide the cabinets, for example, they must also install them and handle all subsequent tasks, such as countertops and finish plumbing, through to the conclusion of the project.

In Chapter 3 I discussed the importance of budget and funding. But even if customers have the necessary funds, they are not necessarily worth your bid. There are individuals who cannot bear to spend $20,000 for a $20,000 job if they can manipulate someone into doing it for less. They call it "the name of the game." A builder friend of mine who has been on the other end of the game calls it "grinding."

The S's (such a lovely couple) typify ardent grinders. Mr. S had recently risen high in his profession, and together with his wife wanted to build a fine new home befitting his new status. They began by agreeing with their architect (a friend) that in return for a reduced fee they would work with a particular general contractor he trusted rather than requiring him to deal with many bidders. But as soon as the S's had the plans in hand, they sent copies to half a dozen contractors, receiving bids clustered in the $200,000 range. Not satisfied, they kept shopping, finally finding a tract-condominium builder who bid $130,000. When he went to work, using his usual slap-it-up techniques, the S's immediately began taking advantage of the architect's loose specs to press for custom work. The Mexican pavers in the foyer constituted a "tripping hazard," and had to be replaced with precisely cast and far more costly Italian tile. The hinges had to be replaced with ones of polished steel to complement other decor. The exterior paint didn't exactly match the sample color chip, so the cost of the paint job should be deducted from the contract price. And so the S's went, grinding the builder down until they no doubt got their house at less than his costs.

Grinders are plentiful among our potential clients. Remodeling consultant Walt Stoeppelwerth says that as a rule of thumb, "One out of five customers is so difficult that the contractor cannot make a profit on the work." Once you are under contract with a grinder, you are stuck in a no-win situation. Even if you get paid every cent in the contract—and more likely you will give away at least a few extras—the client is going to waste a lot of your time. Rather than getting in-

Profile of a Grinder

Grinders sometimes reveal themselves during their first conversations with you. They will soak up as much free consultation as they can when you visit their site, and then phone your office with other questions. Meanwhile, they will let slip that they have already taken four bids. But they want to get even more for even less, so they are going after several others, including yours.

The grinder's greed is particularly striking in contrast to the attitude of fair-minded clients. While the grinder maneuvers for the ultralow bid, fair-minded clients want only confirmation that your bid is within market range. While the grinder squeezes you for additional free services along with the free estimate, the good clients express appreciation for the work you put into an estimate, and limit themselves to getting two or three.

When they show you around their proposed project, grinders classicly minimize the amount of work to be performed. "And over here," they say as they lay out their ideas for a full kitchen and family-room remodel, "we want to do a little something with this space. You know, change the cabinets. Move the sink over a bit. Brighten up the lighting, knock in a few extra outlets. Oh yeah, we'd like to stick in one of those greenhouse windows. Probably ought to pop in some new appliances. And, oh, before I forget, we like that Corian stuff. You could probably get us a builder's special on that. Guess you may as well slap some tile on the floor while you're in here."

When it comes time during the initial visit for you to present your company's way of doing things, the grinders lose interest. They flip through your portfolio as if it were a newspaper advertisement, showing no interest in the carefully built projects and display of thorough management procedure. One veteran construction salesman quickly brings to a close meetings with clients who will not pay close attention to his portfolio. He kept a log of visits with such people and found that they never became customers. So he decided not to waste time with them any longer.

If you do not close the meeting, the grinders likely will, slapping shut your portfolio and asking how quickly you can get them an estimate as they guide you out the door. As you leave, you will likely see another contractor coming up the sidewalk, one of many on the grinders' disassembly line.

volved with such people, you want to spot and disqualify them in the earliest going.

Remember, when you contract for a construction project, you are extending the clients a loan. With each phase of the project, you will be putting in sizable amounts of labor, material and services before collecting your pay. Therefore, qualifying a client should not end with your first visit. To go beyond initial impressions, you can:

• Check references. If you find requesting references awkward, you might try asking clients if they would like to exchange references as you offer yours. Subcontractors, such as a roofer or landscaper, are useful. If the clients have worked with a builder before, inquire why they are not using him or her again.

• Check credit. For about $10, you can purchase a report of a wide range of the clients' financial and legal transactions. (Look in your phone directory for credit-information services.)

• Check for a record of any litigation, especially if you are getting bad vibes. One builder tells the story of feeling suspicious of a client, building for her anyway and later having to sue for his money. He then stopped by the county courthouse and with an hour's research discovered the client had been sued by contractors several times before. If he had put in that hour before signing a contract, he would have saved himself $10,000.

- Request from the client an audited financial statement covering several years. You may find this step especially worthwhile if you plan on signing a major contract with a developer or other client in a high-risk line of work.

As a final step, ask for the clients' response to your references. You want to hear, at the very least, "You are the kind of builder we would like to work with." (Ideally, you will hear, "You are *the* builder we have been looking for.") The clients realize they can get the work done cheaper than you can do it, but they are willing to pay the difference. They are looking not for cheap, but for quality at a fair price.

Though you should be thorough in qualifying clients, take care at the same time not to insist that they pass through an impossibly fine sieve. Don't require them to score a ten on every single test. Bear in mind that some "grinding" is a reasonable response to the high cost of good construction. Remember, the public is conditioned to expect "free estimates" from builders; let the clients pick your brain a bit and count the consultation you give as useful community service.

Be especially careful not to dismiss clients because they happen to be of an allegedly undesirable category of humanity. Such warnings are just stupid bigotry. In my career at one time or other I have been warned by other contractors against doing business with people of virtually every ethnicity and every professional group—I have gone on to have good experiences with them all. I have even had some success working with members of that group most widely decried by builders, namely, architects.

 ## WORKING WITH ARCHITECTS

Builders have good reason to feel angry and frustrated with architects. The response is a reaction to the disdain and even outright hostility with which builders are treated by the architectural profession. One architect summed up the attitude of many others when he suggested, and not entirely tongue-in-cheek, marketing a version of this book for design students under the subtitle "Know Thy Enemy." Historically, architects have sold their services to owners on the grounds that along with superior design, they offered protection against contractors.

Owners frequently do need protection from builders, but they often need just as much protection from architects. Because architects are often trained principally as artists, they sometimes have no feel for construction even after years in practice. At cost control they are notoriously bad. Owners regularly pay them large sums for designs that cannot be built, much less built well, within budget. When they can be, the architects need good builders, who can translate their visions into physical reality.

Nevertheless, as if somehow a building's beauty springs directly from its line drawings, architects rarely credit the contributions of contractors and craftspeople. Their portfolios make no mention of

Gerstel has found that projects built with architects, such as this addition designed by Gene Clements, are among the most challenging and exciting.

builders. Architectural awards or magazine articles featuring architect-designed buildings virtually never mention the builders (even when the builders participated in the development of the project from its inception). The omission is not only arrogant, but puzzling as well, for the implication is that the ability to attract good builders is of little import in an architect's work. But the fact is that poorly built, a design can look crummy. Built well, it can be enchanting.

In their frustration with the egotism and naivete they encounter among architects, many builders elect to avoid them. However, they may well be throwing the baby out with the bath water. For although the architectural profession may have institutionalized certain nasty habits, there are many individual architects who understand construction and respect good builders. Moreover, as Charles Miller, western editor of *Fine Homebuilding* magazine, points out, "Architects have a lot to offer builders. Knowing how to build is important only after you know what to build. Buildings designed by builders are often awful, just terrible. They squander material." Many of the most exciting and challenging construction projects begin with an architect's vision. Architects can also be valuable sources of repeat business. A single architect with whom you establish a solid relationship can bring you as much work as all non-architects in your word-of-mouth network put together.

To work successfully with architects, you need to do two things. First, organize your company to meet their needs. Second, carefully choose the architects you work with. In other words, first qualify yourself to work with architects. Then qualify them at least as thoroughly as other clients.

An architect's initial overriding concern about builders is likely to be this: Will they respect my design? Robert Swatt, a well-known architect based in northern California, says he is put on guard by builders who stop off for a set of plans and hurry away, saying "I'll get you a bid in two weeks," as if they were picking up a piece of meat to put on their scales. What attracts him is "someone I can talk to, who listens, who appreciates that every project is different, instead of glancing at the plans and saying 'Oh yeah, I built ten of these.'"

Once a project is underway, if you want to build an enduring relationship with the architect, you must take care to implement his or her design intentions. Charles Miller has observed that during a project, architects "live in fear that builders will take some shortcut that will sabotage the effect of their details." Builders may think they are doing the sensible, cost-effective thing when they sidestep the details. But from the architect's point of view, the details may be crucial to the integrity of the design. In my experience, on this score the architects are generally right. Frequently, during the course of construction, I have felt that certain details were whimsical or conceited flourishes of the pencil that looked cute on paper but had no real meaning. At completion, I realized those details were the keystones that locked the whole design together.

As a project progresses, communicate relentlessly with the architect. The plans and specs of even very good architects will not be entirely complete. Robert Swatt explains, "It is a dilemma for my profession. We could cut down on omissions by drawing details to one-inch scale. Unfortunately, we rarely have the budget to look at all details in that scale." Therefore, it falls to builder and crew to catch inconsistencies during construction and to bring them to the architect's attention. You can't ask too often for clarification. The architects you want to work with, the ones who will do their job rather than dump the work on you, will appreciate your calls.

Finally, to succeed with architects, you, your crew and your subcontractors must take care not to step between them and the owners. If you spot a problem in the plans, consult the architect. If the owners bring up a problem, refer them to the architect. For if you discuss problems with the owner, the conversation can too easily turn into architect-bashing. And that possibility architects fear. "So much of our work is based on trust," says Robert Swatt, "that the architect is working in the owner's best interest. Nothing can blow that trust more quickly than a contractor saying 'See the way the architect drew this? We could save you a lot of money and do it better.'"

Of course, your own, as well as the architect's, economic interest is at stake here. Architects will not consider you for their work if you get a reputation for undermining them. Inexperienced architects especially can be nervous that some builder will ridicule their construction details and short-circuit their budding reputations. As a case in point, architect Gene Clements recalls on one of his early projects getting stuck with a contractor "who took enormous delight in proving that I was an idiot." Where there was an ambiguity in Clements's specs, the contractor seemed "deliberately to choose the most obnoxious of the possible approaches." When Clements did not specify the finish on exposed ceiling beams, for example, the builder covered them with the same texture used on the drywall. During the same period, Clements worked with a widely respected and successful builder. When that builder found an oversight in the plans, he called Clements, explained that in the past he'd solved the problem a variety of ways and asked which approach Clements would prefer. It's not hard to imagine which of the two contractors Clements invited to bid on the steady stream of projects that came his way in later years.

Playing a supportive role to an architect can be rough on a builder's ego. (Often I sense that the anger contractors express toward architects comes largely from bitterness at loss of turf.) But the fact is that on their projects, architects' decisions take precedence. Architects may not be the "kings of construction," as Clements bemusedly recalls he was taught at the MIT School of Design; at times, buffeted about by owners, engineers, zoning departments and building codes, they may more closely resemble the chief butler. But in that case, on architect-designed projects, we builders are the assistant butlers...up to a point. There are two limits.

The first limit: You do not have to pay for architects' omissions from drawings. Items not included or "reasonably implied" (to use the language of construction contracts) in the drawings are not covered by your bid. In the normal course of events, the owners should pay for these items as part of the legitimate cost of construction. Merely because the owners are finding out about them during and not before construction does not relieve them of responsibility to pay for what they get. A competent architect will forewarn owners that unanticipated items will show up during construction and caution them to keep contingency funds in reserve.

The second limit: You should not pay for an outright mistake by an architect that results in wasted labor or material. The plans may say that you should "verify" dimensions and conditions in the field. You do your best and sometimes even cover an architect's minor error as a prudent investment in good will. But that does not mean that you should allow an architect to hide behind "verify," and stick you with the cost of a major error you did not catch. The classier architects willingly absorb the cost of their mistakes. One architect I respect has half a dozen windows stored in his basement. He specified them for a project, and when they could not be made to work in the design, he paid for them and took them home.

Sometimes you may find yourself with no choice but either to go head-to-head with an architect or to pay for his or her bad work. Twice in my career I have accepted jobs for which the "working" drawings turned out to be anything but. If the projects had been built as drawn, a header would have cantilevered right through a roof. A 6-ft. 9-in. owner would have had to crouch through a 6-ft. 2-in. doorway to get into the master bedroom. A stairway would have ended in a swimming pool. Half a dormer window would have been below the level of the finish floor. And on and on. When called upon for corrections, the architects dropped off hasty sketches as inaccurate as the originals or blatantly attempted to push their work onto my company's shoulders. To keep the projects rolling, I worked out solutions, but I billed the owners. They paid, back-charged the architects and in one case dismissed the architect from the project.

In neither case was I pleased with the outcome. Two potential long-term relationships were wrecked, and likely the architects were denouncing me to their fellows. Much time and energy had been wasted battling with them. I would have been better off avoiding their projects in the first place. Here are four key considerations in qualifying architects, which should help you to separate the good from the bad:

- What kind of work has the architect done? Be alert for architects who have trained in large offices handling commercial or industrial projects and are starting out on their own with small residential remodels. If they nevertheless look promising and worth a bid, make sure you provide in your estimate for the time—likely a lot of time—you and your crew will spend working through glitches in the plans.

- What proportion of the architect's projects get built? A portfolio full of presentation drawings instead of photos of completed buildings provides a tip-off that an architect is putting a lot of projects "in the drawer." If one out of three is getting built, and the architect is taking three to four bids per project, you have only about a 10% chance of getting the job.

- Does the architect know and accept the cost of good construction? Look out for architects who are willing to live with crummy, cheap construction. On the other hand, give some slack on cost awareness. Architects are not estimators, and they are under pressure to be optimistic about cost. One confided, "If we ever told clients what projects really were going to cost, we would never get a commission." Like builders, they are vulnerable to the glamour project. They want to see their most exciting designs built and give in to the hope that those beloved details won't really cost so much to realize in concrete, wood and plaster.

- What type of contract does the architect require? If you have any doubt about an architect's competence, integrity and fairness, think twice before signing an AIA (American Institute of Architects) contract. While protecting architects from liability, AIA contracts give architects enormous leverage over builders, leaving our backsides largely exposed.

To flesh out impressions from a first interview with an architect, check references—not so much from clients, but from other builders who have worked with the architect. I have had a couple of architects get frosty when I suggested we exchange references. But my belief in the value of the practice was confirmed the very first time I used it. Of the three builders I called, the first two became immediately enraged when I mentioned the architect's name. The third was quietly sorrowful. "It was like working for a first-year design student," he said, "and not a very good one at that." I decided to do without that architect's business.

As with other interviews, I prefer to solicit open-ended comment when I call contractors about architects. I get information I never would have thought to ask for, and the conversation usually flows quite naturally across my standard concerns. If not, I ask:

- Does the architect reciprocate the builder's support? "Contractors make mistakes like architects make mistakes," says Robert Swatt. "I don't feel that imperfections, unless they are really serious, should be used to undermine confidence in the job."
- Does the architect become overextended? If so, you probably will find yourself working with hastily recruited, inexperienced staff, and attempting to build from hurriedly produced and inept drawings.
- Is the architect responsive? Do you get the answers you need during construction, and fast?
- Are the architect's drawings clear, complete and accurate? Good drawings allow you to build steadily, with only occasional phone calls to the architect.
- Is the architect fair on change orders? The question is particularly important if you will be working under an AIA contract, for it will give the architect authority over your charges.
- At the conclusion of a project, is the architect prompt and reasonable in making up a punch list?

As a final step in qualifying a new architect, you might consider the long-term prospects. Are there any? Is the architect looking for enduring relationships with builders? Or does he or she churn through them, using someone different on almost every job? Will you always have to produce free bids for the architect's work? Or is there a chance that as mutual respect develops, you may begin to work as a team on projects originating in the architect's network as well as your own?

 # COMPETITIVE BIDDING

Over the years, a large portion, perhaps the great majority, of construction work has been contracted out by means of competitive bidding. Owners distribute copies of the plans and specs to several builders, and choose one on the basis of price, quality or other considerations. I have had severe misgivings about competitive bidding and do little of it. Instead I rely on a form of negotiated bidding I call "price planning" (see pp. 104-110). But for many owners, architects and even contractors, competitive bidding holds strong attractions. As best I can, given my biases, I will discuss competitive bidding's appeals and disadvantages for each group before turning to the alternatives I prefer.

Owners are accustomed to shopping around when they spend a lot of money, and almost any construction project is a big-ticket item. When they shop via competitive bidding, they often can get much free consultation from contractors eager to strut their stuff. Sometimes an owner turns up a real bargain, for example, a good contractor who bids at cost to keep a crew busy.

Competitive bidding also lures architects and designers with the possibility of a low price. They may need it to get their designs built. Or they may take pleasure in getting their customers a "deal." One architect who regularly asks me to bid his projects (and whom I always turn down) has mentioned several times when his projects are complete that the contractor "would never do it for that price again." But though he makes a show of sympathy for the financially wounded builder, his remarks betray an undercurrent of pride at ferreting out the low price. Other architects, although wary of too low a bid, do feel an honest obligation to their clients to seek a tight price via competition.

For some of the builders who like competitive bidding, the appeal seems to be psychological as well as economic. They find it fun, rather like playing high-stakes poker. After winning a low bid they ask, "How much did I leave on the table?" and are jubilant upon hearing that they have played their cards just right, going as high as they could without losing the job. With competitive bidding, if you don't really need the work, you can simply stick in a high markup and wait to see what happens. If you lose, you fold your hand and shrug your shoulders; you're out the time it took you to generate the bid and a few bucks for plans and office supplies. If you win, you can look forward to a large profit without guilt. After all, the owners asked you to undertake the considerable work of estimating with no guarantee of compensation. Quite reasonably, you can generate the bid with your own interest exclusively in mind.

Some builders prefer the relatively "hard-boiled" or "arm's-length" aspect of competitive bidding. (As we shall see, the alternatives imply a much more intense relationship with clients.) Owners ask for a price. You give it. They ask you to do the job or not. If they do, your responsibility is simply to build what is on the plans and in the specs for the price you quote. You're not obligated to warn against problematic existing conditions, to counsel the owner toward more cost-effective results or to offer support services.

For some builders (and occasionally for almost all builders), competitive bidding is the only opportunity to get work. Even where other alternatives exist, it can hold the possibility of more rapid growth. The fastest way to "get big"—to build up a high dollar volume of work—is to pay the lowest wages you can get away with, hire cheap subcontractors, cut corners on projects, bid tight, bid every project you can and build those you get to "industry standard" (which is pitifully low). In my area, one aggressive competitive bidder, despised for his crummy work as well as envied for the ubiquitousness of his job-site signs, has gone from a few hundred thousand dollars annual volume to several tens of millions in a decade.

Competitive bidding does offer to all builders, including those like myself who do it only occasionally, a chance to compare price breakdowns with the other builders involved in a bid. You can see where you are running too high and where too low, and what changes you ought to consider in estimating and project management to keep yourself competitive. If you do no competitive bidding, you run the risk of losing touch with the marketplace. You risk growing sloppy in your cost control, spoiling your crew, loosening the reins on your subs or lazily letting your overhead inflate. If then you suddenly must enter the competitive marketplace, you can't compete.

The drawbacks of competitive bidding for owners, designers and builders run from the merely costly to the nightmarish. For owners, competitive bidding by no means guarantees its one supposed great advantage—a low price. If all the bidders have plenty of work, they may all incorporate a high markup. The owner's "low" bid can then be quite high. One architect figures that seven out of ten times, competitive bidding produces a lower price for construction—which is to say that three times out of ten it does not.

With competitive bidding, owners often do not get a price from the contractors they really want. Those builders land other jobs and bow out of the competition—often to the angry frustration of the owners, who fail to understand that builders do not owe a free bid merely because they kept an iron in the project fire for a time. When bids do come in, owners face the difficult problem of evaluation. A particular model of car, computer or washing machine will be essentially the same product whoever sells it. But a small construction job, however, can be a vastly different product when built by different contractors. From one it's a Chevy that's been in a wreck; from another it's a Lexus straight from the showroom.

Once construction is underway, owners can suffer from the adversarial relationship engendered by the poker game of competitive bidding. They may find themselves battling a change-order artist adept at lowballing the bid, then racking up profits on extras. With worse luck, they select a bidder who can complete the project for the price quoted only by cutting quality to the lowest level that can be jammed past the inspectors—or who cannot complete the project at all. "It's the old money thing," said one architect, who had just watched a contractor fold on one of his projects, leaving behind an array of work that could not pass inspection. "You want to believe the project can be done for the amount of money you have, so you go out on a limb for the cheaper contractor." Then the limb snaps off.

Competitive bidding can also increase the owners' costs for design fees. Not all architects and designers agree, but many feel they must produce more highly detailed plans and specifications for competitively bid projects, to ensure that the contractors are bidding on a like scope of work ("apples and apples"). One large firm in my area, for example, ups its charges from 8% of construction costs to as much as 20% when drawings are being produced for a competitive bid.

Competitive bidding gets ugliest for designers when all bids come in far in excess of budget, as happens with unhappy frequency. The designers find themselves confronted with outraged clients who have paid thousands of dollars for completed working drawings for a project they cannot afford to build. With luck, the designers merely have to go back to the drawing board without additional compensation. At worst, they end up in litigation, facing demands for reimbursement of fees and even for punitive damages for inflicting "emotional distress." To escape the heat, a designer may dig for an extremely low bid, but then runs the risk of having to police the work of an incompetent crew, fight a change-order artist or clean up the mess left by a bankrupted builder.

For builders, competitive bidding can be at once demeaning and exploitive. Sam Clark, a Massachusetts builder, comments acutely in a *Fine Homebuilding* article (August 1984, pp. 61-64), "The fact that highly skilled work (estimating) is provided gratis says a lot about how people in general and builders themselves value a builder's time." Indeed, the public, including designers, has been allowed to become irksomely cavalier about pumping builders to provide free pricing of complicated construction. If you mention the cost of creating that information, you will likely hear in response something along the lines of "Oh well, that just gets absorbed into your overhead." (The last time I heard the phrase was from an architect who, in the next breath, was telling me that he charged for even his initial consultation with clients. He explained that he did not like to give away his valuable time and input for nothing!) For the small-volume builder, the "just overhead" of estimating is typically incurred during long night hours at the office desk. The people who pay are your neglected spouse, kids, friends and—a much overlooked fact—your clients. You generate overhead with all those free bids you do, but only the people who actually hire you pay for it.

Some builders feel the cost of competitive bidding extends far beyond the bidding itself. In the book he wrote after going bankrupt *(Contractor's Survival Manual,* see Resources, p. 223), general contractor William Mitchell urges his readers to find a niche where competitive bidding is not the norm. "For builders who have to survive on low bids," he says, "construction contracting is a real jungle with few survivors." In my opinion, however, Mitchell's warning is particularly germane for small-volume builders. The "jungle" you operate in is peopled not only by toothy predators but also by blundering incompetents.

Frequently your opposing bidders are marginal contractors who are desperate for work, do not understand management or the need to charge for it, slide on insurance and taxes, or just plain can't estimate. Likely, they will soon be out of business. But meanwhile they are stumbling into you, dropping wreckage in your path. The owners who must choose between you and the incompetents are often naive. They may well have never built before and really do not understand that all contractors are not equal, that two people building from the same set of plans can produce a radically different structure. (My own customers are often people who tried to go cheap the first time and learned their lessons.) They are tempted to shut their eyes and embrace that seeming bargain. How do you contend with that? If you lower your standards to compete in price, you, too, will soon be caught up in the destructive cycle of change-order battles, callbacks and lawsuits, all to support poor construction in which you can take no pride.

With all that said, it may be that you can build a solid contracting business by developing an efficient competitive-bidding procedure that wins you a reasonable percentage of bids at a fair markup. There do appear to be people doing just that. But to avoid wasting time on competitive bids you cannot win, do not want to win, or that are for projects that will never be built, you should carefully sift through your opportunities. (Unless your local economy is very slow, you probably will not have trouble finding opportunities to bid. Giving away your time is rarely difficult.) Beyond the steps already covered in the chapters on qualifying projects and clients, you can:

- Ask for the names of the other bidders. Decline to bid against known lowballers and against more than two other builders.

- Determine your chances of prevailing. One competitive bidder with a reputation for doing good work asks owners, "Is there any reason we would not be selected to build for you if our bid comes in within your budget?" If the answer is "No, there is not," the builder assumes he is the preferred bidder. If the owner hems and haws, the builder justifiably wonders whether he is being used. For as one construction consultant points out, "What often happens in a competitive bid is that the owner knows one contractor and is getting estimates from others just to check price."

- As an alternative to determining that you are the preferred bidder, obtain agreement that since all bidders are competent, the low bidder be guaranteed the job. This will reassure you that the owner is not simply using you for purposes of comparison. (In some areas, the European practice of taking the middle bid is reportedly in force, and that's fair, too.)

- As still another option, make a strong play for the projects you do bid. Don't stop at submitting the bid. Precede and follow up the bid with visits to the owners. Present a flow chart (p. 209) of the job. Show your portfolio. Encourage the clients to call your references. Give them a breakdown of your estimate. Ask them to let you know where you are high, especially for the subcontracted work, and offer to go after other subcontractor bids. Offer "value engineering," ideas for cutting price without reducing the project's scope or compromising design. Do everything you can to show the owners that even if you are not low, yours is the best bid. In short, under this option, if you are going to compete, then really compete.

The very best way to respond to a competitive-bid opportunity may be to not bid at all. Instead, before the process gets too far along, you convert the client to a different approach. Frequently clients do not understand either competitive bidding or the options. When an explanation and alternatives are offered, they are often pleased to go another way.

PRICE PLANNING (NEGOTIATED BIDDING)

Among the most appealing options to competitive bidding is "design-build." For owners, there is the convenience of "one-point" responsibility instead of having to bounce between a designer and a contractor. For the design-build firm, there is the advantage of complete control of the project, a chance to make money on all stages of work, not only construction, and increased creative opportunity. Because of such advantages, builders looking for a way out of the competitive jungle often move entirely to design-build. Some cultivate the skills to do the design themselves. Others subcontract design work to architects or designers, or hire them outright.

For almost all general contractors, design-build is often useful for the smaller projects—a deck, a fence and interior renovation. By moving entirely to design-build, however, you do eliminate yourself from that wide range of projects involving independent designers. With a more flexible approach—"negotiated bidding" as it is inappropriately but commonly known—you can become involved in any project and yet stay clear of the competitive morass. For a negotiated bid, you team up with the designer and owner even before the plans are begun. This team approach seems to be making inroads into the old competitive-bid system. The Associated General Contractors (AGC) report a growing recognition among owners of the value of a contractor working with the architect and engineer during design. For the cost of the contractor's participation, says the AGC, the owners get budgetary and scheduling control and development of innovative techniques and procedures equal to many times their outlay. The contractors move beyond competitive bidding and are compensated for their contribution to cost-effective design, most desirably with a contract for the job and/or with a fee.

To my way of thinking, the major flaw in negotiated bidding is its name. It implies that both sides give up something to arrive at an agreement. Although a little horse trading may go on, there is not likely to be that much room for maneuvering. Builders' hard costs for labor, materials and subs are not very flexible if quality is to be maintained. Overhead and risk/profit margins are usually just adequate. What really takes place during negotiated bids is that builders provide input on costs to keep designs within budget. For that reason, in my work, instead of negotiated bid I use the term "price planning." By providing detailed pricing for a project, I assist in planning it so that the owners will end up with a design they can afford to have built. In most cases, I become the builder. If not, I'm paid a fee for my work.

Over the course of several years, I have developed a price-planning process that is popular with my customers, works well for me and has been received enthusiastically by designers and other builders. To begin the process, I set the fee, usually a fixed sum based on an estimate of hours I will spend pricing the job. The fee is intended not so much as a source of income but rather to protect me from exploitation as a free estimating service. With that goal in mind, I set up the fee with several options:

• The fee is waived if I'm awarded a contract to build the job, or if for any reason I must withdraw from the project prior to signing a contract for construction. The waiver provision makes price planning palatable to those clients (the great majority) who have been conditioned to think that contractors are supposed to provide free estimates.

Gerstel's Price-Planning Agreement

Proposal and Agreement for Price Planning

David Gerstel/Builder
Lic. #325650
268 Coventry Rd., Kensington, CA 94707
524-1039

Date: 11/13/89
To: Susan Gold and Richard Peters
Project address: 12 Foggytown Rd.
Project description: Remodel of residence per plans by Michael St. James. Kitchen, Dining room, Deck.
Owners' responsibilities: Providing to contractor up to 10 sets of plans and specs at both preliminary and final stages of pricing.

Contractor's Price Planning to Include:
Initial review of plans and specs by contractor at his office and close review at site together with subcontractors.
One preliminary price for entire project based on preliminary plans.
Up to eight hours discussion and pricing of revisions of preliminary plans prior to production of final plans.
One final price for entire project based on final plans by St. James.
(Both written and preliminary price will be based on detailed and written assumptions, which will be cleared with both owner and architect.)
Presentation to owners (with architect present, if owners wish) of detailed breakdown of both preliminary and final price.

Cost and Conditions of Price Planning
The price-planning services are offered as an alternative method of contracting for construction, and are not offered as a supplement to competitive bidding.

Unless the owner elects to change to competitive bidding for the construction of the project:
   1) A maximum of $1,000 at $40 per hour will be charged for the price-planning services described above.
   2) The charge for the price-planning services will be waived if David Gerstel/Builder contracts for the work.
   3) Additional pricing or planning work beyond the services described above will be charged at $60 per hour and the fee will not be refundable under any circumstances.

If at any time after the completion of the preliminary pricing, the owners convert to competitive bidding, David Gerstel will be paid $60 per hour for all pricing and planning work done to that time. The fee will not be refundable under any circumstances, and this agreement will be void. David Gerstel will be invited by the owner to be one of the competing contractors, and may or may not elect to bid the project.

Contractor's signature _____ Date _____
Client's signature _____ Date _____
Client's signature _____ Date _____

Gerstel revises his price-planning agreement to fit the conditions of each project. Have any agreements that you develop reviewed by an attorney who works with contractors in your area —don't use this one verbatim, in part or in whole.

- If the project falls through — for example, if the customers decide to divorce instead of build, as has twice happened with my price-planned projects — I collect the fee.
- If the clients decide to switch to competitive bidding, I collect the fee and retain the option of joining in the bidding or not. Recently, I have begun stipulating a much higher fee if the clients change to competitive bidding. I thereby discourage clients who really want to use competitive bidding, but see price planning as a way to make sure I bid or to get cheap consultation. I don't want to participate in any hybrid bids. If I'm to do a competitive bid, I want to play my hand unfettered by the greater obligations of price planning.

When I began to develop my price-planning idea, I was not charging enough — in part because of resistance from designers who did not understand the amount of work that goes into a useful estimate. I have upped my fees to a level commensurate with the effort good price planning actually takes. I figure a fee for each project according to the number of hours I estimate I will spend pricing it, not on the basis of a fixed schedule. Typically, the fee runs between 1.5% to 2% of construction cost.

Because clients pay for price planning, and because they are extending much more trust than with an invitation to participate in a competitive bid, I view my obligations to them as much greater. Before signing a price-planning contract, I make full disclosure of my pricing structure. I emphasize that there are no hidden markups. Material is charged at my cost. Subcontractors are charged at cost. Labor (wages plus all labor burden) is charged at cost. I offer to provide copies of written estimates, invoices and payroll records to back my claim. I also show the clients how I calculate my markup for overhead, risk and profit, and what percentage of total price the markup has typically been for similar projects.

I do stress to the prospective price-planning clients that my company is not cheap, for all the reasons that builders committed to doing good work, staying within the law, earning a good living and compensating crew and subcontractors fairly cannot be cheap. But at the same time, I encourage them to compare our price structure to that of other reputable contractors. I want them to feel comfortable that we are well within the fair market range.

If possible, as a last step prior to drawing up a contract for price planning, I evaluate the feasibility of actually building within the clients' budget. I prefer not to charge clients to price plan a job they will not be able to afford. For the many projects that resemble others we have done in the recent past, I can quickly tell whether the budget is adequate to build at least a reasonable version of the design. For less standard jobs, I do a rough written estimate using an abbreviated version of my estimating list. For the most complex projects, I explain to the clients that we may not know feasibility until after the preliminary price-planning work. If it appears the clients won't be able to build even a down-sized version, they can terminate price planning, having wasted far less money than they would have by producing the complete working drawings necessary for a competitive bid.

With full disclosure complete and feasibility determined, I can go to the computer and edit my standard price-planning proposal to fit the project. If the designer has not already been chosen, I make his or her selection the first step. Afterward, design and price planning can proceed in lock step:

- The clients work with the designer to develop preliminary plans. To hold down costs, I urge the designer to keep the drawings simple, providing only enough information to clearly communicate the scope of the work.

- I study the plans and site, and bring in all the necessary subcontractors to price their portions of the project, often with allowances for fixtures, if they have not yet been specified.

- I break down all labor, materials, subcontractor and other costs on my standard estimating form. I try to be liberal in my estimate, feeling that a preliminary price that is a few percents high is more useful than one that is low.

- As I work up the estimate, I record assumptions regarding structural components, architectural detail, allowances for finish and fixtures, and quality level for every phase of the job. (Here my computer is once again a valuable document manager. I can rapidly pump the assumptions into the preformatted form in the computer's memory while I am calculating costs.) I read the assumptions to both the designer and the owner, revising any that are mistaken and changing my estimate accordingly.

- I present the estimate and assumptions to the clients. If the project is over budget, I meet with the clients and designer to discuss possible cuts. Typically, we explore every remotely promising idea for cuts. The clients and designer then shuffle the cuts into a list, beginning with the most tolerable and ending with the least. I price the cuts and send the list back to the clients, who select those they want to make.

- The designer produces full working drawings incorporating the cuts. I do a detailed final estimate.

Price-Planning Sequence

1. Generate a preliminary price from preliminary plans
2. Help develop cuts if needed
3. Price any changes in design
4. Generate the final price from final plans
5. Bill for services if no contract for construction is signed

A Cut List

Johnson House Remodel
Summary of Possible Price-Reducing Cuts
1/25/90
Prepared by J. Johnson

1. Cuts we want to make in any case:
a. Simplify design for downstairs study: reduce cost of built-in bookshelf and omit built-in desk
b. Eliminate glass tub and shower enclosure in upstairs bath
c. Eliminate one window in stairwell
d. Use different (cheaper) window manufacturer
e. Do not move wall in Nathan's bedroom — add new closet in master bedroom without cutting into Nate's room. Eliminate linen closet/chute

2. Cuts we are willing to consider:
a. Cut price on stucco
b. Reduce materials cost on upstairs bathroom mirror
c. Reduce exterior painting costs
d. Eliminate tile floor in storage closet and laundry room
e. Eliminate Sure-Coat drywall
f. Simplify window casing
g. Simplify door casing
h. Eliminate site-built wine cabinet
i. Simplify paint-grade balustrade
j. Reuse existing interior doors
k. Lower contractor overhead and labor allowance
l. Reduce general conditions
m. reduce rough demolition
n. Reduce framing expense
o. Eliminate picture molding downstairs except in family room

3. Cuts we may consider:
Note: These possible cuts are ranked from most desirable to least, because there's a real spectrum of desirability in this category.
a. Tile floor in downstairs bath
b. Eliminate storage slab in crawl space and replace with rough shelving
c. Oak stepping stock for stairway
d. Medium-grade cabinet with marble counter in downstairs bath
e. No drywall work in living and dining rooms
f. Cheaper plumbing
g. Cheaper electrical
h. Sharpen pencil on heating system per discussion with David and Michael
i. Cheaper electrical fixtures
j. Stock exterior doors rather than custom

4. Cuts that start making the job not worth the trouble:
Note: These are not ranked because they're all in the same category.
a. Don't tile kitchen floor.
b. Defer downstairs hardwood floor, replace with carpet or concrete
c. Defer deck
d. Replace upstairs balustrade with half-wall capped with wood
e. Replace downstairs balustrade, screen and cascading stairs with half-wall capped with wood
f. Reduce quality and price of upstairs bathroom vanity
g. Fiberglass shower in downstairs bath
h. Defer all cabinetry in downstairs study

SAMPLE

If a project is over budget, Gerstel gets together with the clients and designer to draw up a list of possible cuts. The clients arrange them in order of preference. Gerstel then prices the cuts. The completed list can be used to guide redesign of the project to budget.

Generally the clients and I are able to proceed from the final pricing to signing a contract for construction. But sometimes more redesign and repricing is necessary. For such additional work I charge an additional fee, typically without possibility of waiver. I have learned that with indecisive clients the design and price-planning process can go on and on, reducing my charge to less than minimum wage if I do not provide for additional fees.

Clearly, price planning requires more work than a competitive bid, but, when it's properly done, the payoff for all parties involved is considerable. For architects and designers, the cooperation of a builder skilled at estimating helps them design for construction that is within the client's budget, so that fewer designs "go in the drawer" or generate lawsuits. Instead of dealing with multiple contractors, as in a competitive bid, they deal with only one. Since they usually recommend or have been recommended by the contractor, or at least know that the owner has chosen the contractor for quality, not cheapness, they know that likely they will enjoy a smooth working relationship. "When you get down to it," says one experienced architect, "the happy jobs have been negotiated. The unhappy ones have all been competitive. In negotiated projects, trust-based relationships can develop, but with competitive bids you're not on the same team."

Unfortunately, price planning is not so unambiguously advantageous for owners. They run the risk of paying more than with competitive bidding. But far worse, they may not get their money's worth. The trust invested in their builder can turn out to be misplaced, with shattering results. I have seen several episodes of "negotiated bids" in which the contractor badly burned the owner. In one case, the contractor withdrew from the project days before construction was to begin, leaving the owner scrambling to find a replacement before the rainy season began and made construction impossible.

More typically, problems arise not from such blatant irresponsibility but as a result of the builders' failure to distinguish sufficiently between competitive and negotiated work. In these cases, the builders have not charged enough for true price planning and as a consequence do not provide service much different than for a competitive bid. When asked to form a team with the owner and designer, they accept the responsibility for an amount equal to a few tenths of a percent, or less, of the likely construction cost. Their preliminary estimates are of a quality commensurate with their pay: quick and rough and often wildly, woefully off—50% to 60% low in several cases I know of and well over 100% in a couple of others. Encouraged by the low price, the owners invest in full working drawings only to find at final pricing that they can not afford to build. They are no better off than if they had gone to competitive bidding in the first place.

With rigorous, fairly compensated price planning done in accordance with the procedure I have outlined, owners should get good value for their money. For an additional 1.5% to 2% percent, they acquire substantial insurance that the 10% to 20% of construction cost they spend with their designer will be for working drawings they can afford to build from. In addition, they get convenience. One builder, who does exclusively price-planned work, says his clients are cash rich and time poor. They do not want to have to deal with multiple contractors, as they must in competitive bidding. With the builder-designer team that comes together for price planning, owners get close to the one-point responsibility available from a design-build firm. They also are assured of a place on the schedule of a good builder, whereas with competitive bidding they often do not even get estimates from their preferred contractors. Finally, as architects I know tell their clients, they are on the right route if they "want high quality and want to take no chances about getting it. With price planning, the builder can put enough in the estimate to do everything right."

For some builders the responsibilities of real price planning can be off-putting. Their operations, or perhaps their personalities, do not accommodate such close working relationships with owners and clients. They prefer to run an efficient competitive-bid mill and let the chips fall where they may. They are certainly right in thinking that price planning can be burdensome. Some of my price-planning efforts have stretched over a full year or more and have at times become quite tense. Nevertheless, to my way of thinking, although it is not for everyone, price planning offers substantial advantages to builders as well as owners and designers.

Price planning also saves time. If you build a company based on price-planned work, you do far less estimating. (You can spend your evenings with your family, not grinding through numbers). The jobs you price, you usually build. You don't deep-six estimate after estimate because projects you bid have been designed irredeemably over budget or because you have been undercut by a contractor too naive to charge for overhead and profit. In the occasional instances when you do not build, you get paid for your estimating work.

With price planning, once a project is underway, you've got less chance of taking a loss on it. You have priced it at least twice and have discussed it in detail with the clients and designer. As a result, you have a thorough knowledge of the project, which mitigates against ruinous estimating oversights. You are not trying to produce a low bid, but a fair price. As a result, you will be less likely to do the wishful estimating that often occurs in competitive bidding.

However, risk is not reduced to zero. A few times I have barely covered my job costs and made nothing toward my own pay, much less a profit, on a price-planned project. The postmortem in each case revealed that I had wanted too badly to be able to perform for the clients' budget. I had shaped my bid accordingly, deviating from my standard estimating procedures.

ESTIMATING AND BIDDING

Setup and Site Inspection
Checklists
Figuring Costs
Marking Up for Overhead
and Profit

SETUP AND SITE INSPECTION

With every project come a few choice windows of opportunity to leap through to financial disaster. The first is the cost estimate and bid. If you have acquired a thorough hands-on knowledge of your trade (see pp. 1-7), you will likely avoid calamity. Sound estimating parallels the typical construction sequence you have worked through again and again in the field, and it is fundamentally a matter of spotting and properly projecting costs for every step in that sequence.

At the start of my contracting career, I made out a new list of steps for estimating each project. Eventually I figured out, as countless other contractors have done before and since, that because the projects in my niche have much in common, I should create a standardized list. I would save labor. I would also gain control, since by working from a standardized list I could avoid overlooking any major items of construction.

These days I begin each estimate by setting up a folder and placing in it a copy of my Estimate "Do" List. One of the first steps on the "Do" list is to obtain a complete set of plans for myself and for every subcontractor. If each sub is given a set, you'll get prices more quickly and easily than if everybody must share one or two sets. Most important, when you give out complete sets, you can reasonably ask that the subs bid all work within their respective trades. You avoid discovering during construction that some aspect of a sub's responsibility was covered only on a page of the plans you neglected to provide—and that you are therefore financially responsible for it.

Once you have plans in hand, add to the folder the following useful items:

- Phone numbers—clients', designer's, building inspector's, etc.—you'll repeatedly need.

Gerstel begins each estimate by setting up a manila folder with all the necessary papers. While the estimate is in progress, he keeps the folder in his sales briefcase. If he signs a contract for the job, he moves the folder to the project-management case and uses it to contain all paperwork until the project is complete. For larger projects with much paperwork, Gerstel substitutes a loose-leaf binder for the folder.

- Sub notes and quotes. For ease of reference, keep your notes of conversations with subs and their quotes together and in the order that the subs will come in during construction. (We are builders, not librarians; we think in terms of construction sequence, not the alphabet.)
- Supplier notes and quotes. Again, order of construction works best.
- Question lists for owners, designer, engineer, building department, etc. I record answers alongside the questions and keep the sheets. In case of later disputes, the information they contain can help provide a resolution. To save my time, and that of the people I'm calling, I consolidate my question-asking into as few phone calls or meetings as possible.
- Assumptions list. If the plans neglect to do so, you should clarify, for example, that the hardwood floor will be tongue-and-groove, not strip; the decking will be fascia grade and not completely clear; the drywall five-point smooth and not four point or textured; the trim custom-milled and not lumber-yard stock.
- Building notes. As you plug away at your estimate, you will hit upon ideas for efficient construction. The good ones—hiring a crane to swing lumber to the top of an uphill site, renting a conveyor to carry dirt out of a basement—will save you substantial labor costs. If you write down your ideas, you don't forget them. One of the fundamental lessons you learn as a builder is that you must write everything down. You handle so much information that you cannot remember all the details, and a single crucial forgotten one can wreak havoc.

Before doing the calculations for an estimate, Gerstel takes off linear and square footages on the plans, using a measuring wheel to speed the work. He notes any conditions not covered on the plans and specs with a dark pencil and marks any special requirements with a highlighting pen.

With your folder organized, you can begin preparing plans and specifications for estimating. I go through both twice, once rapidly for the overall scope of construction and to note subtrades; once to catch finer details. For the second pass, I work my eyes back and forth across the page, slowly dropping down it, as if I were reading closely spaced lines in a book. I hit with a yellow highlighter any unusual, quirky or especially costly details—roof decking to be edge-blocked instead of clipped, joists to be 4x12 on 1-ft. centers instead of 2x12 on 16-in. centers, 30-in. countertops instead of 24-in. As a last step, I write in dimensions and the square footages of wall, floor and ceiling surfaces for later material takeoffs and labor calculations.

Checklist of Site Conditions

Gerstel's checklist reflects his niche—remodeling and building single-family homes. The site checklists of other types of builders will cover different items. Gerstel's list includes:

Access to site
Access to any interior spaces
  to be remodeled
Freedom of movement within site
Parking for crew
Parking for subcontractors
Space for material drops
Space for debris
Power availability
Sanitary facilities availability
Phone availability
Water availability
Excavation
Demolition
Distractions (pets, etc.)
Hazards
Accommodation for owners'
  use of site
Protection requirements, interior
  and exterior
Structural tie-ins at existing
  structures
Finish tie-ins at existing structures

Even for professionally drawn and specified projects, the site inspection offers essential information and reveals the make-or-break costs that don't necessarily show up on the plans. For example, when poor access at a site slows down the movement of material and equipment, much of the workday will go to mobilization instead of construction. When carpenters must crowd into a tight space with subcontractors, their productivity plummets. When staging areas for setup and storage are not available, labor goes into constant reshuffling of tools and suppliers to make elbow room. I learned about the relation of access to productivity during a 10-month rebuild of a complex home on a small lot. At times, with several trades working at once to meet the tight construction schedule, 30 workers crowded onto a site already stuffed with lumber, custom cabinets, windows and moldings. My carpenters, battling their way through the congestion, slowed to half their normal speed.

Hazards at the site—power lines swinging close to a building, open ditches, steep paths—require installation of protective barriers and other safety devices. On remodeling projects, labor and material also go into protecting the building's interior and exterior. Don't overlook the landscaping! Plants killed by construction workers can, like some deceased relatives, acquire beatific qualities, invisible to you as an outsider but heartbreakingly tangible to the survivors, namely your clients. A few years ago on one of my projects, a worker trying to get at the dry rot in an adjacent wall inadvertently chopped down a large rose bush. Through tears, the customer said she'd allow me to compensate her with a substantial renovation of her bathroom. Though her profiteering raised my hackles, her sense of loss was real. After purchasing her house, she had saved that rose bush from strangulation by overgrowth and had nourished it for years. When the designer stepped between us to suggest a settlement of several hundred bucks, I had to go along. The next time I noticed rose bushes at a site, I allowed in my estimate for a few hours of labor and a few sheets of plywood to build protective cages.

During site inspections for remodeling projects, you should pay especially close attention to conditions in existing structures where your new work will tie in. Even capable designers are prone to errors at tie-ins. I have received drawings for a room extension calling for new 2x10 rafters that would have resulted in a 4-in. step down from the 2x6s of the existing roof. For a heavily engineered renovation project, the proposed new beams and posts would have blocked all potential chases for ductwork from the furnace to upstairs bedrooms. On the drawings for a cantilevered second-story addition, the designers failed to note a drop ceiling in the room below, which would conceal their proposed new structural work and greatly reduce building costs.

On a project early in Gerstel's career, a worker wiped out a client's beloved rose bush. The next time roses were encountered on a project, he built a plywood wall around them.

As you inspect a site, note any problems or special conditions on the plans. Write neatly. Arriving back at your office to find you cannot read your own scribbling rouses that urge to rip and throw, activities not conducive to the accurate projection of construction costs. Note problems not only for yourself and your crew, but also for your subs. Although pointing out problems to subs sometimes may result in higher bids, you don't want to let a sub walk into a trap. In the long run, the support and loyalty you get from your subs in return will provide more than adequate compensation.

With my own estimates, I have found it efficient to schedule the subs' site inspections during the same day I make mine. Typically, I arrive at the site right after lunch and invite the subs to come any time during the late afternoon and early evening—those who are working on other jobs can swing by on the way home. For some subs, a site visit isn't essential. But I often ask them to stop by for their plans anyway. Face to face, with the plans in front of us, we can iron out any potential misunderstandings, especially regarding who will handle each of the miscellaneous items that falls between trades. Do the roofers or sheet-metal people flash the roof? Will the plumber or the carpenters set blocks for the shower valves? Does the crew or the stucco sub set scaffolding around the building? Another reason I like subs to come to the job is to meet and chat with the owners. I know that they will leave an impression of competence and integrity, contributing to the owners' trust in my company.

CHECKLISTS

"A construction job is like a lawsuit," I often hear builders grouse, "you can't really know what it will cost until it's over." They are wrong, of course. With a complete, well-organized checklist, a skilled estimator can regularly come within a couple of percent of the actual cost of work shown on good plans. You should not expect, however, to find a ready-made checklist suited to your needs. Niches within the construction industry and companies within the niches are too varied. A generic checklist is no more likely to fit any one company than a generic engine would be likely to fit any one model of car. Two examples of plumbing-bid checklists are shown below.

In developing your checklist, you can, however, get ideas from published material, not only about specific items to include, but also about overall organization. The dominant estimating format for the construction industry is Masterformat, first published by the Construction Specifications Institute in 1963. Masterformat provides for cost estimating under 16 divisions, each of which can be broken down into progressively smaller phases of work coded by number.

With every bid, Jim Lunt includes his checklist (below left) of all those items that could be handled either by Lunt and his plumbers, another sub or the general contractor. Gerstel developed his checklist (below right) for use with all subs after seeing Lunt's.

Two Subcontractor-Bid Checklists

J.W. LUNT
PLUMBING BID/NOTES & ASSUMPTIONS

PROJECT *SANFORD*	PROJECT *11/21/91*
ADDRESS *321 FLORAL DR.*	SHEETS *2*
OAKLAND	
CONTRACTOR *GERSTEL*	
ADDRESS	PHONE

TYPE OF MATERIALS TO BE USED:

	Sewer	Below Grade	Below 1st Floor	Above 1st floor	Vent
DWV	Clay, N.H., ABS, Stub	N.H. (ABS)	N.H. ccDWV (ABS)	N.H. ccDWV ABS	N.H. ccDWV
WATER	(main) L K PVC	(L) K	(L) M	L M	
GAS		wrapped	black galvanized	black galvanized	

THE FOLLOWING ARE/ARE NOT (NIC.) INCLUDED IN THIS BID:

	INC.	NIC.		INC.	NIC.
WATER METER COST AND FEES		✓	SHEETMETAL WORK (any and all)		✓
SEWER PERMIT AND FEES		✓	DEBRIS REMOVAL FROM SITE	✓	
PLUMBING PERMIT FEE	✓		INSTALLATION OF ROOF JACKS	✓	
UTILTY FEES (any and all)		✓	TUB/SHOWER PROTECTION	✓	
WATER PRESSURE REDUCING VALVE	✓		APPLIANCE INSTALLATION		✓
FIXTURES (as spec'd)	✓		SINK CUTOUTS (of countertops)		✓
HOT WATER HEATER(S)	✓		ICEMAKER HOOK-UP		✓
TRENCHING, BACKFILL		✓	NAIL PLATES - provide and install	✓	
CONCRETE DEMO, SAWCUT, BORING		✓	CLEAN-UP OF FIXTURES		✓
PATCHING (any and all)		✓	BLOCKING (any and all)		✓
INSULATION OF WATERLINES		✓	FIRESTOPPING (any and all)		✓
INSULATION OF WATER HEATER		✓	SCAFFOLDING		✓
FOUNDATION, SITE DRAINAGE	✓				

THE FOLLOWING ARE NOT INCLUDED IN THIS BID:
1. HIDDEN CONDITIONS
2. EXISTING CONDITIONS NOT TO CODE
3. STRUCTURAL PROBLEMS

SEE THE OTHER SIDE OF THIS SHEET FOR ADDITIONAL CONDITIONS

Subcontractor's Checklist of Overlap Items

Subcontractor _____ Date _____

	Included in bid	Not included	Not applicable
Permits			
Bill by invoice			
Maintain service			
Protection of surfaces			
Scaffolding			
Demolition			
Trenching			
Broom clean			
Clean fixtures			
Debris removal			
Blocking and firestop			
Cut frame			
Cutouts (sink, plaster, etc.)			
Strap frame			
Cut and patch			
Roof jacks			
Flashing			
Insulation			
Install appliances			
Hook up gas			
Hook up electric			
Write change orders			

Masterformat Divisions

The Masterformat system divides costs into 16 categories, which then are broken into increasingly smaller groupings. Only the main divisions are listed here.

1. General
2. Site work
3. Concrete
4. Masonry
5. Metals
6. Wood and plastics
7. Thermal and moisture protection
8. Doors and windows
9. Finishes
10. Specialties
11. Equipment
12. Furnishings
13. Special construction
14. Conveying systems
15. Mechanical
16. Electrical

Thus, under division six for wood and plastics, you can arrive by progressive subdivisions at 06241.101, for Formica plastic laminates in solid colors.

By adopting Masterformat, you may get some help imposing order on your own estimating procedures, as well as easy use of published estimating information. R.S. Means Company, for example, the most sophisticated publisher of construction-cost information, uses Masterformat codes. If you do, too, you can readily move Means's costs into your own estimates. With Masterformat, you may also be establishing a foundation for future growth, as any professional estimator you may eventually hire will likely be acquainted with the system.

On the other hand, Masterformat may not be relevant to small-volume builders. Because Means's costs books are intended largely for medium- and large-volume contractors, their figures may be of little use to you, especially if your company produces higher-quality workmanship than the pitiful "industry standard." In addition, you may find Masterformat codes cumbersome to use. One capable contractor says he tried them early in his career because he felt "it was the professional thing to do," but "found they were a real pain in the neck."

An estimating checklist organized to fit your operation will likely serve you best. Whatever form of organization you choose, you will probably find it steadily evolving to suit your changing needs. At the beginning of my own career, I instinctively wrote up my estimates to parallel the sequence of construction. Plumbing, electrical, tile and other subtrades appeared in the order that they would actually fit into the flow of carpentry work during a project. Although the decision to follow construction sequence has, with some modification, proved sound, my early checklists had a great weakness. Though they covered each step in construction—foundations, frames, doors, etc.—they often failed to allow adequately for those general costs, such as cleanup, scaffolding and safety, that run throughout a project. When I realized that, I set up a separate division for "general conditions."

At the same time, to increase clarity, I broke carpentry into two divisions, "rough" and "finish." Also, since my projects were becoming more complex and drawing upon more subcontractors (upward of two dozen these days), I created a new division just for subtrade costs. Each of the four cost divisions of my checklist encompasses several pages, with the work broken down into phases and then into individual items. I continue to place phases and items in an order paralleling the construction sequence. I do this not only because as a builder I think about construction as a one-step-after-another process, but also because I have learned that the way in which earlier

Summary Sheet from an Estimating Checklist

```
SUMMARY SHEET

Job address: 121 OCEAN VIEW DR.
Project: MASTER BEDROOM REMODEL/ADDITION
Date: 12/19/91
General conditions:                    $6,286
Rough carpentry work:                   19,743
Finish carpentry work:                  12,425
Subtrades:                               7,915


Contractor's fee:                      $10,100
      Fixed overhead
      Contractor's pay
      Profit
      Profit-sharing/pension plans


Total contractor charges:              $56,469
```

This summary sheet shows the totals of costs figured for general conditions, rough carpentry, finish carpentry and subtrades. A fifth division—contractor's fee—covers the markup for overhead and profit, including Gerstel's pay and his company's profit-sharing and pension plans.

items are handled can influence the construction methods and costs of later items.

So that I can enter costs right on the checklist, each of its pages is divided vertically into columns for labor, material, services (such as tool rental or dumpsters) and subs. The subs column is included in the divisions for rough and finish work usually handled by the carpentry crew, because occasionally a portion of that work is subbed out. For example, on projects involving large amounts of cabinetry and trim, I have found it efficient to contract part of the work to a licensed and insured finish specialist rather than to hire an extra carpenter temporarily.

My estimating checklist has grown steadily to over 20 pages, as I learn of new items and more refined breakdowns from published material, other builders and my own oversights. In this chapter I will present my checklist as one source you might use to develop and organize your own. In Resources (p. 223), you will find other books and journals that can help you to become a skilled estimator. I urge you to draw on all sources. Estimating is like accounting. You can cobble together some backyard system and grope your way forward from it as so many of us in the building industry have done. But since pneumatic tires and high-traction radials have already been invented, why not use them and start out wheeling?

In each of my four cost divisions—general conditions, rough work, finish work and subtrades—most of the phases and items are obvious. But there are some that are invisible, or nearly so. They may not show up on the plans or specifications, or even during a site inspection. During every job, however, you will see material, and especially labor, soaked up by these tasks. Your success as an estimator, and therefore your ability to remain solvent as a builder, greatly depends on your ability to anticipate and allow for the hidden costs.

The first such slippery item on my general-conditions list is "permits," not so much their price, but rather the labor of getting them. There was a time when you could go down to the building department with a few sketches of your project, wait five minutes, pay $10 and be done with it. Today, in urban areas, you may have to work your way through half a dozen or more departments—business license, fire marshal, zoning, engineering, mechanical, electrical and

General Conditions Pages from the Estimating Checklist

GENERAL CONDITIONS

	Labor	Material	Services	Subs

Consult/Permits
Contractor consult
Survey
Plan check
Building permit with license fees
Sewer fee
Street deposit
Parking reserve
Other permits
Bond

Site
Office
Storage shed
Sanitation
Phone
Temporary water
Water charges
Power pole
Electric charges
Material security
Gates and fences
Security personnel
Cover openings
Barricades
Protective canopies
Weather protection:
  Roof tarps
  Doors and windows
  Seal subfloors
  Cover ditches

Travel
Tolls

In the general-conditions division of an estimate, Gerstel figures costs for the phases and items of work that incur cost through all or a major portion of the project.

(continued)	Labor	Material	Services	Subs

Safety
Safety meetings
Safety communication
Rails
Hole covers
Ladder security
Other

Lead organization
Orientation
Order material
Communication with
  Subs, Clients, Designer
  and Contractor

Preparation
Initial setup
Final rollup
Daily setup
Daily rollup
Yard protect
Building protection
Dump

Scaffolding
Site-built
Supplied

Cleanup
Daily
Final

Supervision
(Not included in fee)

Total all general conditions

building—for the initial permit, and then run the gamut again if the project is revised during construction. The process takes hours. You need to charge for your time.

Once a project is underway, the category "weather protection" can easily add hundreds, even thousands, of dollars to a project's cost. In the San Francisco Bay Area, where I operate, the greatest expense is typically the labor and material of securing tarps over open work during the rainy season. In other parts of the country, protection of workers, materials and equipment against cold can be a major cost.

Throughout a project, support tasks such as safety, cleanup and site preparation will consume labor. But over the long run, if these items are neglected, you will pay heavy penalties in crew morale, worker's-compensation premiums and client relations.

During any project, the crew spends much time working at tasks other than the construction called for in the plans and specs, and at the support tasks such as cleanup. Builders allow for these costs in different ways. One I know simply adds 10% to all labor figures in his estimates for dead time or mobilization time. Another figures all workers at $33 an hour rather than his real average of about $23. These practices, however, produce stab-in-the-dark estimates, which indicate that the builders neither understand nor control their costs very well. If you use such rough procedures, at change-order time clients may justifiably resist the imprecise charges and vague margins, and embroil you in costly conflict.

For my estimates, I find it more effective to provide for invisible items not in a single grab-bag category, but as a series of specific line items. Among the most critical is "lead organization time." By this I don't mean the time a lead spends at layout, which is covered along with related items such as wall framing or cabinet installation. Rather, the line is for the time the crew lead spends at management tasks such as initial and daily study of the plans, and communication with the designer, clients, subs, suppliers and myself.

My job-cost records (pp. 68-69) show that lead organization time varies greatly from project to project. On a straightforward two-month project, such as a room addition or kitchen rebuild with good plans, it typically averages six to eight hours weekly, with more hours at the beginning of a project and fewer at the end. For a complex whole-house remodel on which we endure mediocre plans, an unresponsive architect, subcontractors failing to perform and erratic inspectors, fully half the lead's work week goes to organizational tasks. Frequently, in comparing notes on estimating with other contractors, I find that they do have a line for supervision by themselves or their project managers. But they lack one to estimate the hours, and potentially thousands of dollars in cost, a lead must spend on organization.

For a skilled lead such as David Lassman, who can manage a job tightly, the phone is as important a tool as the table saw. The good estimator allows for the time leads use both.

In the rough-work division, a slippery item crops up at the very beginning, with demolition. It's not the basic tearout and removal of foundations, walls, floors, ceilings and interiors by semiskilled laborers that is troublesome—you easily catch that work. But what you can readily overlook, or underestimate by lumping it in with the grosser tearout, is what I call "precise demolition," the skilled separation of one part of an existing structure from another that must remain. Examples are removing 70-year-old plaster from one side of a wall while leaving it unblemished on the other, and removing trim intact so that it can be reinstalled.

Another tricky item to estimate during the rough phases of a project is shoring. You can provide for it under the general-conditions division of your estimating checklist. Or you might simply include it a single time under rough work. But I have learned I have less chance of overlooking shoring if I list it under each phase of rough work for which it could possibly be needed.

Of other recurring items in the rough work, the category "excavation" is the most devilish. Start-up builders especially lose their shirts underestimating excavation. (After all, what's there to digging a ditch or hole in the ground, or scraping away some soil and hauling it away? Answer: A great deal of costly labor when the soil must be chiseled out with a power shovel. Or when it sticks to tools like glue so that they must be repeatedly scraped clean and lubricated. Or when it expands to twice its original volume as you pop it out of the ground.) In creating my estimating checklist, I initially did not even include excavation, vaguely thinking of it as incidental to concrete work. Finally, I stuck it in as a single item. Now it occurs a total of seven times under the site-excavation, concrete, foundation and drainage phases of the rough-work division. I would not be surprised to see it getting even more attention as I continue to develop my checklist.

Yet another slippery item in an estimate, tie-ins, might be of only small importance for new homes. But it can make or break an estimate for remodeling. At every phase of work, the tie-in between old and new work has the potential for absorbing as much labor as all of the related new work. For example, if you add on to an older structure with a humped and sagging floor, framing a smooth transition to the new floor can take more labor than setting all the new joists.

The possible need for margin (an allowance for costs you can't see or even name, but instinctively feel will occur) should likewise be considered repeatedly through the rough work. Frankly, if you end up with much cost recorded on your margin lines, you have a vague estimate. But sometimes you have no choice but to use margin as a buffer against uncertainty. Thus, I have learned to allow margin for projects designed by the flakier architects we work with. While they have an eye for rich detail, they lack the discipline needed to pro-

Rough-Work Estimating Checklist

Gerstel's checklist for rough work, condensed here for purposes of illustration, spreads over five typed pages. He fills in costs for labor, materials, services and subs for each of these categories:

Demolition
Salvage
Rough demolition:
  Bathroom gut
  Kitchen gut
  Other rooms gut
  Finish floors
  Floor frame
  Finish walls
  Wall frame
  Finish ceiling
  Ceiling frame
  Roofs
  Porches
  Doors
  Windows
  Trim
  Other
Precise demolition
Concrete cut
Concrete bore
Clear and grub
Rentals
Dump/recycle
Margin

Excavation
Site
Rental
Shoring
Compaction
Finish grade
Hauling
Margin

Concrete
Stakes (pickup and retrieve)
Layout
Excavate:
  Hand
  Drill
  Backhoe
Form stems
Form walls:
  Chamfer strips
  Decorative strips

Form stairs
Form walkways
Form slabs
Sand
Gravel
Membrane
Rebar:
  Wire
  Tie-ins
Anchor bolts
Pour/spillage
Vibrate
Pump truck
Strip and clean
Point and patch
Haul
Backfill
Rental
Pickups
Deliver
Unload and handle

Waterproof
Apply
Pickups
Deliver

Drainage
Excavate
Gravel
Filter cloth
Drain fabric
Perf pipe
Tie-ins
Deliver
Pickups
Margin

Framing
Nails
Bolts
Hold-downs
Joist hangers
Tie straps
Post bases
Post caps

Other structural fasteners
Crawl vents
Rafter vents
Layout
Structural steel
Wall frame:
  Layout
  Mudsill
  Walls, cripple
  Walls, 2x4
  Walls, 2x6
  Walls, other
  Beams
  Posts
  Headers:
    Door
    Window
  Wall tie-ins
  Plumb and line
Ceiling frame:
  Layout
  Ledgers
  Joists
  Blocking
  Soffits
  Tie-ins
Floor frame:
  Layout
  Ledgers
  Girders
  Beams
  Joists
  Rim joists
  Blocking
  Tie-ins
Roof frame:
  Layout
  Ridge
  Ledgers
  Plates
  Rafters
  Purlins
  Collar ties
  Skylights
  Fascia
  Blocks

Tie-ins
Plywood:
  Shear panel
  Subfloor
  Wall sheath
  Roof sheath
Roof deck:
  2x6
  Other
Furring:
  Drywall
  Plaster detail
  Stucco detail
Blocking:
  Shear wall
  Drywall
  Accessories
  Miscellaneous
Dump/recycle:
  Deliver
  Unload and handle
Rental
Margin

Completion
Corrections
Breakage
Adjust

Modifications
Unique items
Children and pets
Customer participation
Hazards
Access:
  Site
  Rear yard
  Building interior
Cold weather
Hot weather
Rain
Overtime

Total all rough work

duce complete working drawings. A few scrawls on one of their elevations can indicate an elaborate ceiling that will require extensive framing. But that framing will not even be hinted at in the structural details. I allow material and labor for the framing I can see, more for the framing I can imagine, and margin for the framing I know from past experience I am failing to imagine.

Even with the better drawings, I always allow margin for framing lumber, typically about 10% of the board footage I can actually count. I gather from other builders I'm not the only one who regularly runs short of 2x material. I haven't been able to figure out where it goes. Maybe over the fence into neighbors' weekend projects. Maybe to that fifth dimension I once saw defined as the place where car keys go after evaporating from your pocket and where socks come to rest after disappearing from the wash.

A good portion of that extra framing lumber, however, probably goes into miscellaneous "furring and blocking," another of the rough items you can lose sight of during estimating. I first became aware of this years ago, when I worked with a team of framers on a condo complex. Back in the thicket of studs and plates I noticed an old-timer, the "pick-up man," whose sole job it was to complete our frames. When I went on my own, I found that in remodeling, the furring and blocking for bath accessories, waste lines, offsets and the like require an even larger portion of the labor than it does in new construction. I now allow for it in six different lines.

The items under modifications, at the end of the rough division, do not consume material or labor in and of themselves, but can increase the cost of construction for any previous item. Pets, for example, can distract your crew every step of the way through a project. One builder tells of a client's mutt constantly grabbing her crews' tools—what doggy fun!—and depositing them in the swimming pool. The happy hours her crew spent diving for their equipment showed up on their time cards under rough and finish phases of work.

A more serious modification is for "access." If a site is tight, or construction is distant from the parking and off-loading areas, labor can increase dramatically. Unfortunately, the effect of access on productivity is almost impossible to estimate precisely. As a rule of thumb, estimate high to protect yourself. For some jobs you will find it prudent to plug in an amount for labor equal to 10% or 50% or even 100% of the total labor for all other phases. Your crew can easily be slowed to half their normal speed if they must constantly struggle against tight and steep paths, congested rooms, high staging and scaffolding and heavy subcontractor traffic.

Finish-Work Estimating Checklist

Gerstel's finish-work checklist covers items typically handled by his crews. As with general conditions and rough work, his actual pages provide a space for costs for labor, material, services and subs.

Exterior finish
Felt/Tyvek
Siding
Siding tie-ins
Corner boards
Rafter tails
Rafter blocks
Rafter molding
Fascia
Fascia tie-ins
Knee braces
Door casing
Window casing
Skirts
Treads and risers
Stair rails
Pickups
Delivery
Unload and handle
Margin

Windows
Windows
Skylights
Greenhouse windows
Additional hardware
Head layout
Caulking
Flashing
Prep and prime
Pickups
Deliver
Unload and handle

Doors
Doors
Jambs
Hinges
Prehang and bore
Hang
Single
Double
Rehang in jamb
Locksets
Deadbolts
Sills
Threshold
Combo

Closers
Weatherstrip
Plates
Kick
Push
Pickups
Deliver
Unload and handle
Prep
Prime

Interior stairs
Carriages
Skirts
Risers
Treads
Newels
Posts
Rails
Balusters
Wall rails
Pickups
Unload and handle
Deliver

Interior trim
Nails
Hardware
Windowsills
Window stool
Extensions
Window apron
Window jambs
Window case
Door case
Base:
    Pieces
    Inside joints
    Outside joints
    Footage
Wall caps
Stair skirts
Picture molding:
    Pieces
    Inside joints
    Outside joints
    Footage

Crown molding:
    Pieces
    Inside joints
    Outside joints
    Footage
Columns
Closet shelves/poles
Paneling/wainscoting
Mirror trim
Light valances
Tie-ins
Pickups
Deliver
Unload and handle
Margin

Accessories
Paper holder
Towel bars
Glass shelves
Grab bars
Shower rod
Medicine cabinet
Toilet-paper holder
Deliver
Unload and handle

Cabinets
Site measure
Inspect and correct
Built-ins
Vanities
Kitchen uppers
Kitchen base
Kitchen full height
Window seats
Doors only
Pulls
Tie-ins
Pickups
Deliver
Unload and handle

Decks
Nails
Hardware
Piers
Other foundation

Ledgers
Joists and rims
Decking
Posts
Rails
Balusters
Pickups
Unload and handle
Deliver

Fences
Posts
Rails
Screens
Pickups
Deliver

Gates

Completion
Caulk
Epoxy
Drywall touchup
Hardware adjust
Door adjust
Cabinet adjust
Minor breakage
Minor corrections
Returns
Callbacks
Customer instructions

Modifications
Unique items
Hazards
Children
Pets
Customer participation
Access:
    Site
    Rear
    Interior
Overtime

Dump/recycle

Total all finish work

Subtrade Estimating Checklist

Unless the volume is very small, Gerstel puts out speciality work to subcontractors, the ranks of which continue to swell under the impetus of new laws, new technology and new fashions.

Asbestos	Rails	**Paving**	Corian	**Acoustical ceiling**
	Security	**Insulation**	Granite	
Masonry	Other	Wall	Plastic laminate	**Wallpaper**
		Floor		
Sewer	**Plumbing**	Ceiling	**Floors**	**Landscape**
		Weatherstrip	Hardwood	
Heating	**Sprinkler**	Caulk	Carpet	**Paint**
		Water heater	Resilient	Interior
Gutter and spouts	**Electrical**			Exterior
	Phone	**Drywall**	**Glazing**	Cabinet finish
Other sheet metal	Cable	Stock	Mirrors	
Flashing		Hang	Window lights	**Dump/recycle**
Fireplace	**Special systems**	Scrap	Other	Delivery
Hood, fan, ducts	Intercom	Tape		Pickups
Bath fan, ducts	Alarm	Cleanup	**Garage doors**	Unload and handle
Vents				Totals
	Roofing	**Tile**	**Tub and shower**	
Metal work			enclose	
Ladders	**Stucco**	**Countertops**		**Total all subtrades**
Stairs		Ply base	**Appliances**	

Finally, under the rough division, you need overtime for tightly scheduled projects. Be generous in estimating overtime. Remember, crews working overtime are tired and don't produce at their usual speed. If tasks are done at a wage rate 150% of normal, then when you factor in slower production, those tasks may cost twice as much as they would if done during regular hours.

Many of the hidden items that arise in the finish-work division (see p. 123) are repeats of those already discussed under the rough division. Again you see tie-ins, margin and modifications, including access and overtime. At cabinets, however, there's a new wrinkle—inspect and correct. One cabinetmaker candidly told me, "There are 2,000 measurements made for a typical set of kitchen cabinets. I'm going to make some errors." The frequency of errors and the labor needed to correct them vary hugely from cabinetmaker to cabinetmaker. I have found a high figure especially necessary with modular cabinets shipped from distant factories, the worst being certain highly advertised European units sold out of kitchen boutiques. When boxes or scribes or kicks are wrong, getting the right one takes weeks (if you're lucky) when your manufacturer is halfway around the globe and the local rep is someone who was selling water purifiers or lingerie six months earlier. My best local cabinetmakers beat the socks off the modular companies, and not just for reliability, but for quality and price, too.

Toward the end of the finish division come 11 items under completion. This category is especially important at the end of finish, when many small adjustments, much touchup, repairs of minor breakage and so on are needed. Sometimes these items are discovered by the client only after the crew has left the project and must be taken care of as a callback. Early in our careers we may think of the callback and other completion items as avoidable errors that we should not charge for. But no one does perfect work. Completion is an inevitable and essential part of the project—crucial to cinching client satisfaction and word-of-mouth references. You must do it, and you must therefore provide for it in your estimate.

The work itemized in my subtrade division typically is done by subcontractors. I include columns for crew labor and material because if there is only a small amount of specialty work—say, a single room to drywall or a small counter to tile—I often find it more cost effective to do the work in-house than to sub it out. Also, when the trade is subbed out, crew labor and company material is nevertheless required. Sometimes the costs are quite direct, as when a carpenter spends a day altering the frame for the sheet-metal sub's ductwork. Some of the costs are invisible, as when the plumber arrives at a small kitchen remodel and the carpenters lose productivity keeping out of the way.

One item entirely missing from my estimating checklist is labor burden—costs such as worker's-compensation insurance that go on top of wages (pp. 45-47). I don't cover it separately as a line item, because it is included with labor costs entered throughout the estimate. Thus, if I am paying carpenters $18 an hour and labor burden is 50%, I estimate carpentry labor on each item at $27 an hour. Other estimators enter only wages as line items and enter burden as a single total for all labor in a project. There is no single right way to estimate. Only one rule holds for all approaches: Don't leave anything out. As builder John Ward says, "A guess, even a bad guess, is better than leaving it out. You get badly hurt not when you guess, but when you forget to guess."

 # FIGURING COSTS

To succeed as an estimator, you must hold tight to this crucial distinction: Figuring the costs for a project and entering them on your estimating checklist is one procedure (I'll discuss it in this chapter); bidding a project is another separate procedure (I'll look at it in the next chapter). A bid, of course, is based on an estimate of costs, but it also includes markup beyond costs for fixed overhead, profit and risk. With markup, you have flexibility. Just how much is determined by your economic position and your relationship to the client. Your markup can legitimately be modified for strategic or even subjective reasons. For example, you might bid for the renovation of a building on the historic registry with only enough markup to cover your overhead. You'll take your "profit" in prestige, publicity and pride of work instead of money. On the other hand, you might double your usual markup for a competitive bid offered to certain clients. You don't really need the work, and you figure if you are going to build for these characters you had better get well paid for it.

But prior to marking up and bidding, when you are simply figuring your costs for the project, you must free yourself as much as possible from subjective considerations. You do not want a project's glamour, a hope for large profits or hunger for work to interfere with your objectivity. Your aim in figuring hard costs is not to get the job, but to account in your estimate for every item in the project.

Accuracy and efficiency in figuring costs require clear procedures and techniques for:

- Calculating material for each item to be built by workers on your payroll.
- Calculating labor for each item to be built by your workers or yourself, if you perform hands-on work at the site.
- Taking bids for services and subtrade work.
- Checking and totaling all the material, labor, service and subcontractor costs entered on the checklist.

When you enter costs on the checklist, save yourself the trouble of putting in dollar signs—they just clutter the page. Also, round off all figures to the nearest dollar. One remodeling manual recommends leaving the pennies in to impress the customer with the thoroughness of your estimate. But that's just a cheap trick, and good builders don't need cheap tricks.

Figuring material costs is usually the easier part of producing an estimate. To figure material quantities, you need only count units, or do a few basic arithmetic or geometry computations, then modify the results for actual field conditions. Among the key modifications are these:

- For volumes: When figuring excavation of soil, allow for expansion. In my own area, soils typically expand by about 50% as they are excavated. Thus, if I figured 40 cu. yd. to be excavated for a foundation, I would count on moving 60 yd. (40 x 150% = 60). Similarly, when figuring the volume of material to put into a project, allow a margin for spillage and other waste. Thus, when figuring concrete for a foundation, add 10% to the amount yielded by the formula.

- For areas: When you figure 4x8 sheet material—plywood, drywall, etc.—you usually need to add whole sheets for any leftover area under 32 sq. ft. To prevent waste with tongue-and-groove material, such as plywood for a subfloor, at the end of a run you can substitute blocking for the tongue and groove rather than using up a whole sheet to get a narrow tongued ripping. When figuring material for sheathing, drywall or siding, don't deduct for door and window openings. The material cut away is usually scrap.

- For framing: For joists, the formula will give you the number of joists that will be installed for a given length of floor or ceiling. But you also need to add for rims and other doubles, fireblocking and culls. For studs, as the formula suggests, by figuring one per foot you will generally include enough extra material to provide for doors and windows, corners and partitions in an average wall. When it comes to figuring material for roof gables, there are two schools of thought. Some builders look at a gable as two triangles that equal a rectangle and figure enough material to frame it, as shown below. Others feel there is so much waste in framing a gable that it is best to figure as if the gable were a rectangle with a height of its longest stud. I favor the latter method. Figuring gable ends "solid" contributes to the overall margin needed for framing lumber on a job.

For framing larger structures, such as big additions or new homes, instead of figuring numbers of sheets or sticks of lumber, calculate material quantities in board feet. Estimating in board footage does have disadvantages. It does not necessarily give you the exact lengths of the lumber you will need to build. You or your lead may have to figure again for that information if you contract for the project. Also, you may get inexact quotes from your lumberyard, since the price of material often varies according to length. The attractive aspect of fig-

Fundamental Builders Formulas

To succeed as a builder, you must understand elementary geometry. If you don't, take the necessary courses.

Volume for excavation:

$$V = w \times d \times l \times \%$$

Width x depth x length x percentage for expansion

Volume for concrete:

$$V = w \times d \times l \times \%$$

Width x depth x length x percentage for spillage

4x8 sheet material:

$$S = w \times l \div 32$$

Number of sheets = area (width x length) to be covered ÷ area of sheet (4 x 8 = 32

Number of studs in a wall laid out 16 in. on center:

$$S = 1$$

Number of studs = length of wall in feet

Joists in floor laid out 16 in. on center:

$$J = [(r \times 3) \div 4] + 1$$

Number of joists = run of floor x 3 joists for every 4 ft. plus final rim joist and any doubles

Framing a Gable

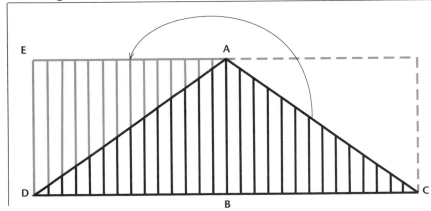

If you cut a gable end into two triangles (ABC and ABD) and place one on top of the other, you get a rectangle (ABDE). Cutting a long stud for one of the triangles gives you a short stud for the other.

Formula for Board Feet

Lumberyards usually quote larger quantities in board feet, so by using board-feet formulas, you can quickly figure the cost of material in a project.

$$\frac{\text{Width (in.) x depth (in.) x length (ft.) x number of boards}}{12} = \text{board feet}$$

Examples:
Board feet in one 8-ft. 2x4:

$$\frac{\text{2 in. x 4 in. x 8 ft.}}{12} = 5.33 \text{ board feet}$$

Board feet in a unit of 300 10-ft. 2x6s:

$$\frac{\text{2 in. x 6 in. x 10 ft. x 300}}{12} = 3,000 \text{ board feet}$$

Number (N) of 12-ft. 2x4s in 1,000 board feet:

$$\frac{2 \times 4 \times 12 \times N}{12} = 1,000 \text{ board feet}$$

$$N = \frac{12 \times 1,000}{2 \times 4 \times 12}$$

$$N = 125 \text{ boards}$$

Cost (C) of 80 8-ft. 2x8s at $480 per 1,000 board feet:

$$\frac{2 \times 8 \times 8 \times 80}{12} = 853.33 \text{ board feet}$$

$$C = \frac{480}{1,000} \times 853.33 = \$409.60$$

uring in board feet is that with the use of formulas you can figure your lumber quantities for larger structures with remarkable speed. Thus, to get the board footage of joists, you can multiply the floor area by a factor corresponding to the size of the joist. For example, if the joists are 2x12, you can multiply the area by 1.65 and get the board footage including a 10% waste factor. For a 20-ft. by 40-ft. two-story home engineered for 2x12 joists, you simply multiply 2 x 20 x 40 x 1.65, to get 2,640 board feet). It is beyond the scope of this book to cover all the board-footage formulas. For an extensive and clear presentation, I recommend Paul Thomas's *Estimating Tables for Homebuilding* (see Resources, p. 223).

Once you have figured the quantities of materials for a project, you need prices from your suppliers. For efficiency, you may want to give your complete materials lists to your suppliers and have them provide you with an itemized price. I prefer to get unit prices over the phone, even though it means figuring totals myself. My suppliers employ highly skilled salespeople, who often make valuable suggestions for alternate materials during our conversations. If I didn't make phone contact, those suggestions probably wouldn't surface.

Since material prices change so rapidly, small-volume builders who bid only intermittently will probably find it necessary to get new ones for every estimate (or group of estimates). Always find out how long the prices will be good for. If the suppliers cannot hold their prices until you will need the material, include margin for potential inflation in your bid. If there's a likelihood of a steep rise in price, simply include allowances in your bid, with any savings or additional cost to accrue to the client. Do not underestimate the danger of inflation. During the 1970s oil crisis and ensuing upward ratcheting of the consumer price index, rising material costs wiped out the entire margin for profit and overhead for a new home built by a contractor I know.

You enter the real danger zone of estimating when you move to figuring labor costs. When builders make a hurtin' mistake in an estimate, low labor figures are usually to blame. With material, you are counting things you can see on the plans. Unless you overlook an item, make a mistake in your math or are surprised by increased prices, you will get your costs about right. With labor, however, mere counting and proper math do not guarantee accuracy. First of all, you've got the problem of those "invisible" items. Second, you are forecasting the behavior of that notoriously unpredictable creature, the human being. Sometimes you are predicting the output of workers without even knowing who the workers will be. Output varies immensely, by as much as 1,000% according to one expert, between workers with nominally the same skill classification. And the output of any single worker can vary greatly from project to project, depending on his or her physical or psychological condition. For example, the productivity of an excellent carpenter who once worked for me dropped sharply when his relationship with his lover soured. After they made peace, his productivity jumped up again.

As discussed in "Job Costing" (p. 65-70), to sharpen your predictions of labor productivity, you need records of past performance. You must know how long carpenters have previously taken to form a foot of single-story foundation, to frame a foot of standard-height wall, to install a sheet of subfloor, to hang a door, to case a window. In a pinch, you can get your numbers from published manuals of labor cost or from other builders. Or you can reconstruct them from your own past experience as a tradesperson. But there is no substitute for the fully detailed written record from your past jobs, both those on which you performed the hands-on work personally and those built by your crew.

With such records, you can estimate labor for each item on your checklist in three steps: First, look up the unit cost from the past project that most nearly resembles the new one. Second, if necessary, adjust the unit cost up or down to fit the new project. (To make your adjustment, consider especially the relative productivity of the crews for the old and new project.) Third, multiply the adjusted unit cost by the quantity of the item in the new project. Finally, multiply that result by your labor rate.

Labor burden for typical small-volume contractor

If the contractor whose burden is itemized here pays a carpenter $20/hr., his cost for the carpenter's labor is $29.30/hr. ($20 x 146.5%).

FICA	7.15%
FUTA (approximate)*	1.00%
State unemployment (approximate)	4.50%
State disability	1.20%
Worker's comp	17.90%
Liability insurance	4.80%
Crew supplies (variable overhead)	10.00%
	46.5%

*Note that as the year goes on, labor burden will decline for permanent employees, since you pay unemployment taxes only on the first $7,000 of their wages.

A Phone Quote

Gerstel uses a checklist to take quotes from well-known and trusted subs, and then only for small jobs. Otherwise, he relies on written estimates.

Subcontractor (supplier)
License number
Liability and worker's comp
Person taking quote
Date
Quote
Length of time need for work
Who will do
Description of work covered by quote
Exclusions

Was description of work and exclusions read back to subcontractor?

Did the subcontractor verify that description of work and exclusions as read back was correct?

Signature of person taking quote

For example, you are projecting labor cost for a one-story foundation 60 ft. long. Your cost record for such a foundation built 18 months ago shows three hours of labor per foot, inclusive of all work: layout, excavation, forming, rebar installation, placing and vibrating concrete, stripping, cleanup and pointing. You plan to use the same crew on the new job, but in the year and a half since the last job they have built several other types of foundations and increased their skills. Also, your record is for a job done in the winter, whereas the new project will be built in the summer, when your workers are always more productive. You figure, conservatively, that labor time will drop to 2.75 hours per foot, and that the labor rate for your three-person crew will average $20 per hour. Therefore, labor cost for the foundation will be $3,300 (60 ft. x 2.75 hr./ft. x $20/hr.).

Regardless of how you figure labor costs, you must be sure to include in them not only employee wages but also the labor burden. From our discussion of the spreadsheet in Part 3, you are already acquainted with the items included in labor burden—namely, all those insurance and tax costs that you incur on top of wages for every hour of labor that goes into a project. The percentages that builders add to wages to cover labor burden vary greatly. For example, one builder adds only 33%. His remodeling company employs non-union crews who receive few benefits, and he accounts for crew supplies—all the material and small tools that crews use up on a project—not as labor burden but as part of fixed overhead. He has an excellent safety program, has experienced few injuries and thereby has reduced his worker's-compensation premiums to minimal levels. Another builder adds 100%. Her workers are all union members, and her labor burden includes the union's medical and dental plans, vacation pay and pension plan. Her employees perform dangerous structural-repair jobs, and several have been injured, driving up her worker's comp rates. A third builder, who may be the most typical of small-volume builders, operates a non-union crew and does both remodeling and new construction. His labor burden is broken down in the illustration above left.

Entering the costs for subtrades in your estimate can be the easiest and safest of estimating tasks. Typically those trades—plumbing, electrical, sheet-metal, tile, etc.—will be handled by subcontractors. They will figure material and labor costs and markup, and assume the risk of error. Nevertheless, to ensure reliable figures, observe the following guidelines:

• Do not personally estimate costs for trades you do not know well, or for which you have no reliable unit costs. (Sometimes subs will give you unit costs you can directly apply for their work. A plumber, for example, might tell you to allow a fixed amount per fixture for labor and material).

- Require that subs' bids cover all work shown on the plans and specs for their trade unless explicitly deleted. Do not accept bids that itemize tasks, for the subs can then pass on the cost of any oversights to you, claiming their bids did not include the item.
- Make sure that all work incidental to a trade is covered either by the sub's bid or elsewhere in your estimate. For example, for the trenching, removal of old pipe and blocking incidental to plumbing, verify either that the plumbers have it covered or that you have allowed crew labor for it.
- Get subcontractor bids in writing. Rely on phone quotes only if you have worked with the sub repeatedly, and if the work is relatively small in scope. I learned my lesson here when I understood a sub to tell me $1,400 on the phone, and he thought he was saying $4,200. We discovered our miscommunication only after my bid was in and accepted. Fortunately for us both, at the last moment the clients canceled the project.

Your estimate for material, labor and subcontractor costs on a project is not complete until you have reviewed it carefully. After completing your estimate, set it aside for a day or two before picking it up for a fresh look. At that time:

- Check for omissions. Review the plans, specs and estimating checklist to make certain you haven't overlooked anything. Keep a special lookout for hidden costs, such as lead organization and access modification, discussed in the previous chapter. Check that in figuring your labor costs, you've allowed for any raises that will occur during the project.
- Look at high-cost items to be handled by your crew, whether they involve a single expensive unit, such as an eyebrow dormer, or multiple low-cost units, such as framing straps and hold-downs. Check your material cost entries against the quotes from your supplier. Make sure you have correctly multiplied both labor and material unit costs by the number of units in the project.
- Check your entries for subcontractor work against their phone quotes or written bids. Make certain that all the incidental costs are covered for each trade.

For my estimates, once I have completed my review and am satisfied that the entries on my checklist are as accurate as I can make them, I move the totals to my summary page. Now I am ready to figure my markup for overhead and profit and to incorporate my cost estimate into a bid.

Overall Labor Check

Gerstel likes to see how many crew weeks of labor he has allowed and whether the allowance corresponds to his records of labor used on similar past projects. For example, for a complete kitchen remodel, Gerstel would proceed as follows:

1. Decide that a lead and apprentice, assisted by temporary workers during demolition, will do all carpentry and some subtrade work, such as drywall.

2. Figure the cost for the lead and apprentice including labor burden will be $1,680 a week and the temporary agency will charge approximately $100 a day for laborers.

3. Total the labor cost for all items on the estimate and come up with $14,208.

4. Determine that six person-days ($600) of temporary labor are needed for demolition and miscellaneous tasks.

5. Figure that $13,608 ($14,208 − $600) is left for the lead and apprentice.

6. Divide $13,608 by $1,680 to get 8.1 weeks (i.e., enough money has been allowed in the estimate for just over eight weeks of work by the lead and apprentice).

7. Check the job-cost records to see that in the past, kitchen remodels with a scope of work similar to the new job have taken between seven and nine weeks. The labor estimate, with an allowance of 8.1 weeks, therefore looks reasonable.

MARKING UP FOR OVERHEAD AND PROFIT

If estimating for labor is the riskiest aspect of pricing projects, marking them up for overhead and profit is the aspect about which the most confusion reigns. Builders often are wrong about which items to include in overhead and which not, uncertain about the legitimacy of profit and confused about how to figure markup for both.

The difficulties may arise from conflicting views purveyed by industry authorities. If you look through construction manuals, you will find one expert suggesting an 8% to 10% markup to cover overhead. Another tells you to figure it at 15%. Still another insists on a minimum of 50% for overhead and profit combined, and pushes hard for 75%, with the lion's share going to overhead. If you ask around among your more seasoned contractor buddies, you may find equally puzzling spreads. In the San Francisco Bay Area, small-volume builders typically mark up "10 and 10" (10% overhead, 10% profit). But others, having operations seemingly no different, have decided on 25% or 30%. One high-profile company boasts of markups around 70%. Meanwhile, the chief estimator for a very large and highly respected firm reports that his company builds on a 4% to 6% margin for overhead and profit combined.

Underlying the contradiction is an explanation: This huge spread in markups occurs for a reason—the industry experts are not just off their rockers. Rather, each of them (though they may forget to say so and write or speak as if they were addressing the whole industry) is really talking about one niche in the vast construction world. One person's experience is with the construction of medium-sized commercial and public facilities. Another person's is with urban office buildings. One is talking about small remodeling projects (such as bathroom renovations and window replacement). Another has in mind that more extensive kind of remodeling that comes close to being new construction.

Though it can be bewildering, the variety of advice does underscore an important point. Markup—just like keeping the books and estimating—is really a highly individual matter. You can get thought-provoking suggestions from many experts, but you cannot automatically apply their formulas to your jobs any more than you could expect to walk around comfortably in another contractor's boots. You must, so to speak, find the shoe that fits, then wear it. You must design your markup to fit the particular configuration of your company.

To develop workable markups for your projects, first identify the costs you should include under fixed overhead. Many of those listed at right are obvious. Others deserve explanation:

- Liability insurance, health plans and pensions. You may find it most accurate to include these costs in labor burden (p. 53-55), but if you do not, be sure to include them in fixed overhead.
- Shop and office. If you rent your shop or office, you will include the payments in your disbursements journal and easily spot them as a fixed cost. But if the office and shop are in your home, you won't be writing a check for their use (unless you incorporate, and rent your company the space). Nevertheless, you should assign the office and shop a market value for the purpose of figuring overhead. For example, my shop and office are in my home, and no cost for their use shows up on my books. But I still plug $600 "rent" into my calculation of monthly overhead. Otherwise, I would be providing my clients with the use of the space gratis.
- Cost of working capital and assets. The money you have in your business account and tied up in your truck, tools and office equipment could be invested elsewhere and bring in dividends or interest. The income you lose is known as the "cost of money." One business adviser suggests figuring the cost of money at four points above the inflation rate. As an example, if you have $20,000 in working capital and $10,000 in equipment while inflation is running at 6%, your cost of money is $250 a month:

$20,000 + $10,000 = $30,000; 6% + 4% = 10%
$30,000 x 10% = $3,000; $3,000 ÷ 12 months = $250/mo.

- Bookkeeping and payroll services. If you hire a bookkeeper and/or payroll service, the cost will appear in your disbursements journal. But if you do these tasks yourself, assign your work a market value and include it in overhead.
- All management work, including sales, estimating, contract writing, project management and office management. As a small-volume builder, quite likely you will perform these tasks yourself. Again, you need to assign your work a market value. If you are not yet skilled at management, you may want to value your work modestly. Thereby you keep your overhead low and your company strongly competitive, as a start-up operation generally needs to be. But as you gain experience, you should push the value of your work up to a level commensurate with the salary of a professional estimator and project manager.

Frequently, inexperienced builders, especially those still working with the tools, do not include the value of their office and field-management work in overhead. They give it away, or view it as something they must do to get themselves a job at their trade. The

Fixed Overhead Costs

A first step in figuring out appropriate markup is to make a list summarizing fixed-overhead costs. Gerstel's list includes:

Liability insurance
Health plans
Pensions
Truck purchase and maintenance
Shop
Equipment depreciation
Office
Office equipment
Phone
Postage
Office supplies
Office heat and power
Legal services
Bookkeeping
Payroll services
Accountant
Licenses
Taxes
Bad debts
Dues
Subscriptions
Entertainment
Other promotion
Advertising
Working capital (interest loss)
Assets in company (investment income lost)
Interest on bank loans
Secretarial services
Sales
Estimating and bidding
Contract negotiation
Project supervision
Office management

omission leads them into other errors. You often hear them say that they underestimated a project and merely "broke even" on it. But if you dig deeper, you find that in fact they were able only to cover their labor, material and subcontractor costs. Therefore, they did not even break even. They lost money. Along with their other overhead expenses, they lost the pay they should have received for managing their companies.

A similar naivete also marks the comments about profit you hear from start-up and even fairly experienced builders. They will speak of making a "good profit" on a job. But in fact, they made only enough to pay themselves wages for their hands-on work at the job site and/or a modest salary for their managerial work. Neither the wages nor the management pay is profit; the first is compensation for a labor cost and the second is part of overhead. True profit is income beyond job costs and overhead.

It seems to be voguish in recent business literature to say "Profit is not a dirty word." In fact, it has become a dirty word for good reason. Much gouging and exploitation goes on in the name of making a profit. On the other hand, a reasonable profit is not only legitimate but also necessary. Your company must have profit to build up working capital and a reserve against hard times. You need it to provide your workers with a profit-sharing program (p. 188-190), one of the best means to building stable and reliable crews. Most important, you need profit as a margin against the enormous risks of construction contracting. You take these risks with every job you do.

One builder who does high-risk structural repair work marks up bids 10% to 30% beyond costs and fixed overhead. But she counts on only 1.5% staying in her company. The rest she figures may well go back out to cover the losses occasioned by her work.

Risks

Contracting can be a risky business, and you need profit as a hedge against these risks. Listed below are some of the more common risks builders may take with every job:

Fire
Theft
Storm
Flood
Earthquake
Volcanic eruption
Heat wave
Illness among key people
Injury to key people
Strikes
Material shortages
Failure of subs to perform
Client exploitation or harassment
Inflation
Capital shortage
Structural collapses
Worker errors
Lawsuits

Having convinced yourself of the need for overhead and profit, you must still determine how and how much to mark up your bids. In a few pages I will describe my own method. But traditionally, small-volume builders figure overhead as a fixed percentage of costs based on the experience of the previous year. Thus you can:

1. Figure your job costs for the previous year (the total of all the cost columns in your disbursements journal, excluding fixed overhead); for example: $423,678.

2. Total your fixed overhead costs, including the value of your management work and any other costs that don't appear in your disbursements journal; for example: $49,352.

3. Figure the percentage you would have needed to mark up your costs to cover overhead; for example: $423,678 x ?% = $49,352. ? = 11.64846%. (Proof: $423,678 x 11.64846% = $49,352.)

4. For convenience, round off to 12%. In the current year you mark up your costs on each project by 12% to cover overhead.

Even with the traditional approach, allowing for profit and risk is not so mathematical an operation as allowing for overhead. Profit is really a goal, albeit a necessary one, and setting a reasonable goal is a matter of judgment. Your judgment may be influenced by such factors as what you think the market will bear (usually not a great deal, in the highly competitive construction world), your sense of what is fair to charge your customers, and your sense of the risks involved. My advice: Be fair to customers, but do not underestimate the risk.

Once you have determined a percentage for both overhead and profit, you may choose to mark up your costs by a single figure intended to cover them both. But for greater accuracy, treat overhead and profit as two distinct considerations, and mark up in two steps. First, mark up your estimated costs for overhead. Then mark up the total of your costs and your overhead for profit. To illustrate, in the case of a 20% versus a 10 and 10 markup on $8,000 in job costs:

A. $8,000 x 20% = $1,600 overhead and profit
$8,000 + $1,600 = $9,600 total bid

B. $8,000 x 10% = $800 overhead
$8,000 + $800 = $8,800 job costs plus overhead cost
$8,800 x 10% = $880 profit
$8,800 + $ 880 = $9,680 total bid

Determining markup by means of fixed percentages can work reasonably well if you stay in a narrow niche—building custom homes, or remodeling kitchens and baths, or installing foundations. Then your job costs and overhead are likely to remain in a fairly stable relationship, so that a consistent markup consistently produces the needed charges. Many small-volume builders, however, do a great variety of projects, partly because it's more fun, partly because that is the work that comes their way. Your repertoire can run from decks and fences to structural repair work; from small remodels to large additions to new buildings. The ratio of overhead to costs for labor, material and sub work varies greatly. So does the time the projects take to complete and the level of risk they entail. The percentage of markup you need for one type of project may be very different than for another. As you can see in the example at right, if you mark them up the same percent, you can do nicely on one type but poorly on another.

A more dangerous potential pitfall inherent in the percentage method of markup is this: The percentage is figured as a ratio of the previous year's job costs to overhead. But the current year may bear little resemblance to the previous year. Costs can change radically from year to year with relatively little accompanying change in overhead, as shown on p. 136. When that happens, if you use a percent-

Consequences of Using Same Percentage Markup on Different Types of Projects

Shown are the results of using the same percentage of markup on an eight-week, high-risk, $20,000 repair job, and an eight-week, moderate-risk, $100,000 high-end kitchen remodel. In one case, the allowance for overhead and profit is miserable; but in the other case, it's good.

	Repair Job	Remodel
Costs		
Material	$4,012	$21,962
Labor	$15,119	$37,312
Services	$1,023	$4,109
Subs	0	$29,617
Total	$20,154	$93,000
20% markup	$4,030	$18,600
Projected income per week for overhead, profit and risk	$504	$2,325

Consequences of Using a Percentage Derived from a Previous Year for the Current Year's Markup

In a big year, one elaborate project involving expensive subcontractors could double your usual receipts. Meanwhile, fixed overhead for bookkeeping, office equipment, phone, etc., may hardly vary. Therefore, if you use a percentage of markup for overhead derived from your figures for the big year, in the smaller year you are going to be way short of what you need to cover overhead. In the example below, you would be over $14,000 short of the markup you needed.

	1989	1990
Gross receipts	$773,000	$411,000
Overhead	$34,000	$31,000
Job costs	$671,000	$323,000
Markup needed to cover overhead	5.1%	9.6%
Shortfall if 1989 percentage is used to figure overhead in 1990		$14,527

age from a previous year in the current year, you may be severely undercharging—or overcharging. You can come up short of funds to pay yourself for managing your company and for your other fixed-overhead costs, or you can push your bids too high to be competitive.

Even people who use the percentage method of figuring markup don't seem very satisfied with it. A highly respected builder in my area who relies on the percentage method to figure overhead admits that the method is little better than a wild guess at a reasonable markup.

An entirely different approach to markup—one based on the duration of a project, not a percentage of its costs—can be more effective. For my work, I have developed just such an approach. While it will not fit every construction company (no system ever does), it may be of value to many small-volume builders. Here's how it works:

When I mark up the cost of a project for a bid, rather than invoke a percentage, I concern myself primarily with the length of time the project will take to build. I figure that my company keeps two crews working steadily (although the same principles would apply if we had one crew or three, or if I were working as an independent tradesperson). We book work for those crews and only those crews; because of our commitment to quality and reliability, we do not book a project and then find a crew to build it. Therefore, the number of weeks in a year our crews have available to build defines and limits our earning power. We must make a certain amount above cost for each crew week to meet annual needs for overhead and profit. To figure that weekly amount:

- I calculate how many crew weeks of work we are likely to produce during the year. For example, allowing for vacation time, illness, bad weather and some slack time between projects, I estimate that each crew will work 44 weeks a year. With two crews, that's 88 crew weeks available annually.
- Next, I set minimum annual goals for overhead and risk/profit. For example, I might set $16,000 for fixed overhead for office rent, supplies, postage, phone, etc.; $50,000 for additional overhead (largely my earnings for sales, estimating, office management and project management); and $22,000 for risk/profit. The total annual minimum goal for overhead and profit is thus $88,000.
- I determine how much markup we need to move satisfactorily toward this goal. If we expect to get in 88 crew weeks of work a year and we aim to take in a minimum of $88,000 annually for overhead and profit, we need to make at least $1,000 per crew per week for markup. We need to make the $1,000 a week whether the project will take four weeks or nine, whether it is large or small, whether it will be done entirely by the crew or involve many subcontractors.

As you may have discerned, the time method of markup can result in quite different markup percentages for different types of projects—and that is just as it should be. For example, if a crew does a small renovation job involving no subs and requiring $15,000 in labor and material, and it is expected to take six weeks, following the procedure above you would mark up a minimum of $6,000. That's 40%. Does that seem exorbitant? It is still well below the level of markup urged by one national remodeling consultant.

By contrast, you may do an elaborate custom home involving $321,000 in labor, material and subs, which is expected to take 36 weeks. A $1,000 per week markup—$36,000—would be 11.2%. Depending on your area and the state of the economy, that may or may not be unduly competitive.

I do want to emphasize that the $1,000 per crew week is an example of a projected minimum. You may have cause to go up from your minimum under a variety of conditions:

- Difficult construction or a large volume of construction that greatly increases your project-management responsibilities. In contrast to the small renovation project, you cannot put a crew on the job and check in briefly once or twice a week. Rather, you must spend long hours at the site and constantly be in touch with your lead, the owners, subs and designer.
- Market rates. If the time method of markup gives you a percentage unreasonably below market rates, you might choose to bring it in line. Thus, you might determine the 11.2% markup in the previous example to be too low, and go up to $1,500 a week, resulting in a 16.8% markup.
- Need for higher financial rewards in absence of other rewards, such as execution of an interesting design or construction of a socially valuable project.
- High risk of problems during or after a project.
- Clients who are likely to be extremely demanding and/or interfering during the project.
- A competitive bid for a project you do not need, but would take if there were a high profit margin.

On the other hand, you may at times have to reduce your desired minimum markup, as in the following situations:

- Need to provide work for yourself or your crews during a slow spot in your schedule.
- Desire to do a challenging or socially useful project for which the client's budget can cover your overhead, but not provide much for risk and profit.
- Opportunity to do a project with someone who seems to be a promising long-term connection. But take care here! You work

cheap once and a client will probably want you in the future only if you are cheap again. A perverse dynamic can be at play. You do premium work cheap, then when you try to charge an appropriate price, the client may think, "Hey, if we're going to pay a premium price, let's get a top guy. This is a cheap guy."

Marking up on the basis of time as opposed to percentage is not the ultimate solution to the problem of allowing for overhead and profit in bids. For a small-volume builder, however, the time basis does have advantages. It allows you to define your annual goals for overhead and profit, and then to mark up a wide range of individual jobs with those goals clearly in mind.

The time-markup method can also be adapted to different kinds of operations. My company does a mixture of new construction and re-modeling work, but a builder I know who specializes in labor-intensive repair and renovation jobs uses a similar approach. First, he sets a minimum goal for overhead and profit. Then he asks the people on his crew how much they want to make and, therefore, how many hours they want to put in during the year. He totals up their anticipated hours. Then he figures how much he must add for profit and overhead to each hour of labor to reach his goal. For example, his workers might want a total of 8,000 hours, while he has projected a goal of $88,000 for profit and overhead. So he will add $11 for each hour of labor he expects to put into a project. During the year, he regularly reviews his progress toward his goal, adjusting his markup as necessary.

Whatever method of marking up a bid you use, once you have figured your costs, added your markup and set your bid, stick to it. If your bid is too high for a client, you can offer "value engineering," the development of alternate specs for the project that will cut costs while retaining the essentials of the design.

Perhaps you can seek alternate sub bids if some seem unnecessarily high. But do not, as frequently happens to start-up contractors, let a client or designer talk down your cost estimate or markup. And do not make anxious last-minute cuts on a well-reasoned bid because you fear you might lose the job. Often, the night before submitting a bid, I have been tempted to shave it. I worry that I might be high and that I will lose the project. Often enough I do get the job. If not, that's probably just as well. For there is much truth in the old builder's saying, "The best bid you will ever make will be the one you don't win."

CONTRACTS

PART 6

Why and What Kind
The Agreement
Conditions
Change Orders
Subcontracts

WHY AND WHAT KIND

At the outset of our careers as builders, many of us harbor the delusion that working on a handshake is somehow classier than using a written contract. We tout our informal agreements as a sign of our ability to inspire trust in our clients and to do business by the good old values of trust and honor instead of having to rely on legalistic paperwork. Unfortunately, our disdain for the written contract is sentimental balderdash and sometimes merely a rationalization for laziness. Thorough written contracts are a must. Not only do they give you something to fall back on during times of disagreement, but they also can prevent disagreements—very costly ones. The most likely reason for a lawsuit developing out of a project is not some screwup during construction. Instead, it is the failure to spell out in advance what is to be built, for how much and under what conditions. Providing a thorough contract is a service you owe your clients, as well as a favor you do yourself.

Contracts can be tedious. Occasionally though, as you wade through the subject and discover some clause that will save your hide should you ever again get into awful situation type X, Y or Z, you can feel a surge of joyful relief. There's even a bright side to the boredom. Contracts need be tedious only a single time. Once you have studied the subject and created your contract, you can use it over and over with only an occasional update, simply filling in a few blank spaces for each project. Count yourself lucky if you find contracts boring. It's when you don't take the trouble to create a good contract and as a result end up in court, draining off your working capital to cover an attorney's fees, that the subject becomes irresistible.

The Elements in a Contract

A contract is first and foremost a tool for communicating with clients. The contract that communicates fully includes five elements: the plans and specifications from the designer or architect, any clarifying assumptions you must add, and your agreement and conditions.

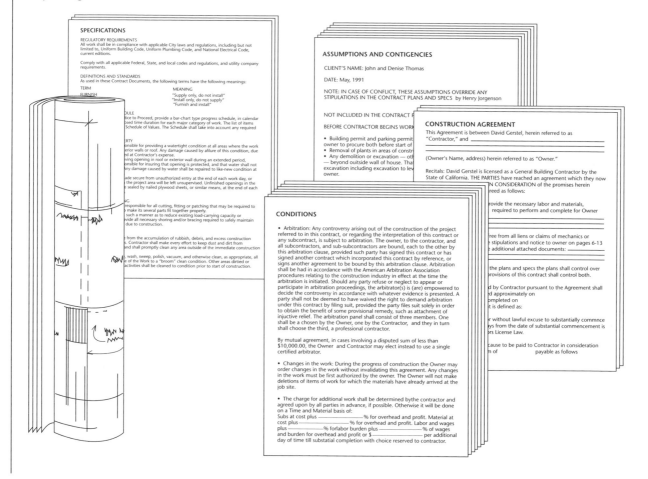

Contracts are of two basic types. Under the "lump-sum" contract, the client pays you to do the work described in the plans and specs for a fixed amount of money. With "time-and-materials" (T&M) contracts, on the other hand, the client reimburses you at cost for labor, materials, subcontractors and incidental items, and also pays you a fee for overhead and profit. Over the years, several ways of providing for that fee have evolved. Each represents a response to a weakness in a preceding arrangement.

Under one early arrangement, owners and contractors settled on a percentage of total costs as the fee. Such "cost plus a fee as a percentage" contracts have fallen into disfavor. Because the more expensive a project becomes, the more a builder makes, owners feel these contracts do not give builders any direct economic incentive to control costs.

As a result, the "cost plus a fixed fee" contract evolved. Again, the owner reimburses the builder for costs, but the fee for overhead and profit is fixed at some agreed-upon amount, typically a fair-market percentage of the estimate for the project. Even if a project runs over the estimate, the fee remains static. Consequently, there is no incentive to let costs soar.

But neither is there any incentive to control costs. For that reason, a third option arose, the "cost plus a fixed fee with a guaranteed maximum" contract. Here, again, you collect a fixed fee for overhead and profit. But you also agree to charge for costs only up to an agreed-upon maximum. If you go over it, you eat the excess.

Guaranteed-maximum arrangements offer security to clients, but they put builders in a lose-lose situation. If the builder exceeds the maximum, he or she takes the loss. If costs stay below the maximum, the clients benefit, but the builder gains nothing for the risk of accepting a maximum. As an antidote, there is a fourth (and final) variety on the T&M contract, which might be called "cost plus a fixed fee with guaranteed maximum and split difference." Here, along with setting a fixed fee and a maximum cost, you and the owner agree to split any savings on the guaranteed maximum. For example, if the guaranteed maximum is $110,000 and the project comes in at $100,000 inclusive of all costs and your fee, the owner pays you half of the $10,000 savings, or $5,000.

For all the options that have been invented, there is still no perfect construction contract. (Thus, even with the guaranteed maximum refinement, an unscrupulous builder can inflate the top figure to boost profit. Or a greedy owner can manipulate a hungry contractor into accepting an inadequate maximum.) Nor are there any hard-and-fast rules to determine the right type of contract for a particular project. In good measure, the choice boils down to your and the client's preference. Personally, I prefer working under lump-sum contracts. They involve less paperwork. You can simply include a schedule of fixed payments in the contract and bill accordingly, rather than laboriously tallying all labor and material every couple of weeks. Lump-sum contracts are clean and clear. You take the risk—if your estimate is high, you enjoy the gravy; if it is low, you suffer the loss. Your client is not burdened with risk, and that is as it should be, especially for the typical customers of the small-volume builder, who are often naive about construction. They have enough to worry about without sweating out the accuracy of your price.

T&M contracts, by contrast, build tension into a project (or, to put it another way, they provoke the client to mess around in your business). Given the high cost of labor, clients naturally worry that your crew may be moving at less than optimal speed, and may begin to spy on them. When that happens, workers can become edgy or angry, and slower at their work. Customers may also want to fight with you over which costs are reimbursable and which should be covered by your fee. They may balk at paying such normal job costs such as for removing waste, correction of minor errors, legally required break time and crew safety meetings.

For example, on my first T&M project, when the bill came due, the client complained about the "terrible excess" of material I had left over. To her untrained eye, the three full sheets and pile of scraps remaining from the 1,000 sq. ft. of drywall I had hung and taped in her home did look like substantial waste. To me, it seemed like evidence of an accurate takeoff, but she demanded I deduct its value from her bill.

With T&M contracts, you can also find yourself at loggerheads with the owner over additional fees for changes in the work, especially upgrades in finishes. For example, in a kitchen remodel, the countertops may be changed from plastic laminate to granite. The clients think they are merely asking that you tell the countertop sub to change materials and, therefore, that you deserve no greater fee. You realize that the upgrade involves much more complex coordination with the subs and liability for an $8,000 instead of a $1,200 installation. You feel you must ask for a commensurate increase in overhead and profit. In my work, I have felt it fair to charge for upgrade change orders using a markup one-half to two-thirds of the markup used for entirely new elements in the project. Some clients have begrudged even that, while others have been appreciative.

Although it is possible to provide for the equitable settling of potential disputes (see pp. 144-152) in a T&M contract, the basic sources of tension remain. The clients worry over crew efficiency and waste of material, and resent paying for corrections and an increased fee for change orders. You and your crew grow irritable at their distrust, and you tire of defending your charges. Because of the potential for such dispiriting conflicts, it seems to me that T&M contracts should be used sparingly and only in instances like these:

- Early in your career, when you take on a project you're not confident you can accurately estimate. Actually, you should turn down such a job (see "Qualifying Projects," pp. 85-89). But if you must take one on, as we have all been tempted to do when we were starting out and work was sparse, a T&M contract can protect you from ruinous losses. To compensate your client for picking up the risk, you can offer an attractively low markup. At the start of my career, I typically worked on such jobs for wages plus 5% of costs to cover my out-of-pocket overhead. The jobs came out okay, with my customers getting decent work at a low price and myself gaining valuable experience. In retrospect, though, I know I was lucky. Had my estimates been badly off, I could have created a financial disaster for both myself and my clients.

- The client insists on a T&M arrangement. One of my customers does. As a vice president of the world's largest construction company, he is habituated to T&M arrangements, which are used on all his projects (Saudi Arabian airports, Korean power plants, etc.). He chooses T&M for his personal projects as well because he feels it offers the greatest flexibility and that as long as my crew and I work steadily, he gets a fair price.

- A project needs to go on a "fast track"; that is, it must start before drawings are complete and before a firm price can be established. For example, clients of mine needed to begin their house before the winter rains, but the architect was late completing the plans. He had produced enough for a building permit, but not nearly enough detail for a firm bid. We agreed to start immediately, with the architect providing necessary detailed drawings as work progressed and the owner paying us costs plus a fixed fee.

- A lump-sum price is meaningless. For example, you are contracting for work on a severely deteriorated building. Clearly the hidden costs you will encounter once you open the walls will exceed the cost of the work you can see and accurately estimate.

When Gerstel inspected this old house, he discovered raccoons roaming the attic, decks sagging like great hammocks, fissures running from gutters to foundation, retaining walls sliding downhill and turrets tipping away from the main structure. He explained to the owners that there would be so many hidden costs in the renovation that a lump-sum contract would be meaningless, and he could work only under a time-and-materials agreement.

In the course of your career, you are likely to have some projects for which lump-sum contracts are appropriate and others that will best be done under T&M contracts. As with other documentation and procedures—company policies, estimating and bidding—you must develop contracts suited to the particulars of your own operation. No stationery store or mail-order catalog will offer a ready-made contract exactly matching your needs. On the other hand, help is available from many sources, and in a few steps you should be able to produce excellent documents:

1. Study the following chapters in this section to introduce yourself to the two basic components of a construction contract—the

agreement and the conditions—and also that crucial extension of the contract, the change order.

2. Collect a variety of contracts. Among those you should look at are the documents from the American Institute of Architects and, especially, the Associated General Contractors. (There is probably an AIA chapter near you. See Resources, p. 223, for the AGC address.) Off-the-shelf contracts from mail-order houses, while sketchy overall, often will contain several useful clauses. If you're lucky, a seasoned builder in a niche similar to the one you occupy will share his or her contract with you.

3. Take all the contracts and cut them up into their various clauses. Group the clauses that cover similar issues.

4. Study the clauses and use them as a basis for creating your own. Don't hesitate to write your own clauses in your own language. Strive to be thorough and clear rather than to sound like a lawyer.

5. Write up any additional clauses you think you need.

6. Arrange all the clauses in a logical order to form an agreement and a set of conditions.

7. Take your draft to your attorney for any needed adjustments and corrections.

Alternatively, if you're lucky, you may find a local builders' group that has created a contract similar to what you need. You can then modify and improve it by drawing on other contracts and your understanding of the needs of your operation. Either way, be prepared to put time and money, including fees for attorney review, into creating your contract. If you take the lazy way out and try to get by with a mail-order contract, you will probably leave gaping holes in your understandings with your clients. Down the line, you will pay the price.

THE AGREEMENT

Both the agreement and the conditions of a construction contract must treat all parties fairly. You should provide for the division of obligations on every issue that can arise during construction. But at the same time, you want to avoid draping a thick blanket of protection over yourself while tossing your client a mere fig leaf. A one-sided contract may end up defeating its selfish intent, for contracts are required by law to be reasonably evenhanded. It can be bad for business, too, putting off clients who already have been educated to distrust contractors. On the other hand, a fair contract can increase clients' trust in you and serve as an effective sales tool.

The agreement for a balanced construction contract should cover four basic concerns:
- The parties to the agreement
- The work to be performed under the agreement
- The schedule on which the work is to be performed
- Payment for the work

Typically, the parties to an agreement are the builder and the clients. That is the case even with AIA contracts for construction (see pp. 93-99).

The agreement may contain within itself a complete description of the work for small projects, such as installing new windows in a home, building a deck or building partitions in a store. But for larger jobs, such as a residential addition, restaurant remodel or new home, it is not possible to describe the project in a few lines. Instead, the agreement "incorporates by reference" the plans, specifications and any modifying or clarifying addenda such as your assumptions. They in turn provide the direct description of the work.

One addendum I especially value is that which I add to AIA contracts. These contracts so heavily favor the architect, that I feel I must override some of their provisions and add protective clauses from my own conditions. If you work under AIA contracts and want to develop modifications to protect yourself, make sure you have them reviewed by an attorney familiar with the AIA documents, so you don't create contradictions that might cripple the contract.

An agreement should be exact in its reference to the plans, specs, addenda and any other documents it incorporates by reference. It should name the person who executed each document, state what project it is for, list the number of pages it contains and give its date.

Typically, agreements refer to the schedule for a project in terms of "substantial commencement" and "substantial completion." Both terms should be described precisely. Thus, substantial commencement could be defined as the beginning of demolition or excavation and substantial completion as the date the project can be used by the client for its basic purposes. A kitchen, for example, could be considered substantially complete even though a few electrical fixtures were back-ordered. Note that a project is "substantially complete" before it is "complete." One construction attorney warns against ever using the term "final completion." Owners may seize upon this phrase to demand additional polish on a project as an excuse for withholding payment.

Construction Agreement

David Gerstel/Builder
B-1 Lic. #325 650
268 Coventry Rd., Kensington, CA 94707

This agreement is between David Gerstel, herein referred to as "Contractor"
and _____
(Owner's name, address, phone number), herein referred to as "Owner."

Recitals: David Gerstel is licensed as a General Building Contractor by the State of California. The PARTIES have reached an agreement which they now wish to reduce to writing. IN CONSIDERATION of the promises herein contained, it is mutually agreed as follows:

• Contractor promises to furnish the necessary labor and materials, including tools, required to perform and complete for Owner (project description) _____
_____
_____

in workmanlike manner, free from all liens or claims of mechanics or materialmen, subject to the stipulations on pages 5-12 ("Conditions") and other provisions on the additional attached documents _____
_____

In case of conflict between the plans and specs the plans shall control over the specifications and the provisions of this contract shall control both.

• The work to be performed by Contractor pursuant to the Agreement and shall be substantially commenced approximately on _____ and shall be substantially completed on_____
Substantial commencement is defined as _____
_____

PAGE 1

• The failure of the contractor without lawful excuse to substantially commence work within twenty (20) days from the date of substantial commencement is a violation of the Contractors License Law.
• Owner promises to pay or cause to be paid to Contractor in consideration for his performance the sum of _____
payable as follows _____
_____

• Performance of the work covered by this Agreement, including a General Arbitration Clause, is subject to the stipulations stated on the following pages 3 to 11 and the Notice to Owners on pages 12 and 13. Owner and Contractor understand and agree to the stipulations and Notice and include them as part of this agreement.
• The Owner has a right to rescind this contract for a period of three days after signing and to reclaim any deposit made at signing.
The offer to build for the sum stipulated in this agreement is good for_____ days or longer if the Contractor wishes to extend the offer.
IN WITNESS THEREOF, the parties have executed this agreement and Owner hereby acknowledged receipt of the following Documents and Notices.
1) One (1) Copy of the required "Notice to Owner."
2) One (1) Copy of this Agreement properly signed and all attachments.

Executed on this _____day of
_____ 19_____
Contractor _____
Owner(s)_____

PAGE 2

Gerstel composed his 13-page agreement (two pages of which are shown above) with the help of attorneys to suit his particular operation. Develop your own agreements with the help of an attorney familiar with the contractors' laws in your area.

Payment is the most complicated issue covered in an agreement. Usually, you first collect a down payment to hold a place in your schedule (your state may limit the amount). For the remainder of the payments, the agreement must provide a schedule. One school of thought recommends loading up the front end as heavily as possible—for example, collecting 50% at the start of a kitchen re-model and 40% more when cabinets arrive. The practice, however, may be illegal in your state, and at best it is likely to alienate clients. Given the reputation of contractors for leaving work incomplete once they have their money, clients may become so suspicious of one who proposes a front-loaded schedule that they seek another builder.

A fair schedule provides for payment to be made promptly after work is completed, and in one of three ways: If you are working under a T&M agreement, you simply collect for any costs and proportionate amounts of your fee as you build. Under a lump-sum agreement, you collect a series of fixed payments provided for in the contract (see the example shown above). Alternatively, you can collect a series of payments corresponding to the percentage of work

Assumptions

Gerstel generally includes in contracts a list of assumptions to clarify or correct the plans and specs.

Assumptions and Contingencies

David Gerstel/Builder, Lic. #325 650

Client's name: Bryant
Date: 4/90
NOTE: In case of conflict, these Assumptions override any stipulations in the plans and specs by John King dated 3/14/90, 3 pp.

Not included
• Building permit
• Asbestos removal
• Reframing of hallway ceiling (scope of work and cost to be determined after finish ceiling is removed)
• Plumbing fixtures (owner has already purchased)
• Closing hardware for casement windows

Allowances
• Electrical fixtures: $1,300
• Weatherstripping for exterior door: $100 labor and material and/or sub
• Exterior door lock and deadbolt: $200
• Tile: $389 (could be low)

Regarding demolition
No removal of hardwood flooring in breakfast nook or hallway. Vinyl tile and underlayment will be removed in kitchen (Note: If VT contains asbestos, then owners will have it removed by asbestos contractor and general contractor will give owners $200 credit).

Regarding drainage
Tie-in of new downspout assumes that drain line is ABS plastic and not more than two feet below grade.

Regarding framing
No new subfloor will be installed. Existing subfloor will not be renailed except on a change-order basis.

Regarding heating and other mechanicals
Upstairs register will be left in existing location. New ductwork will run through existing or new framing shown on plans. If a register has to be relocated or additional chases built, the work will be done on a change-order basis.

Regarding other sheet metal
Les Williams will mount hood fan on existing chimney with necessary fittings and connect to hood fan with ductwork run through existing chase or utilize existing ductwork. No reconstruction of duct chase is included in price.

Regarding plumbing
The contractor is not responsible for failure of the plumbing fixtures provided by the owner.

Regarding electrical
All electrical fixtures will be selected by the owner but purchased by the electrical contractor at his cost plus 15%. Chime to be reusable. Gut existing fuse box in closet and replace with new subpanel. Replace or shield Romex wiring in existing furnace-laundry area and mount laundry outlets.

Regarding windows
Windows will be paint grade from 4th Street Woodworking to match existing. Owners will provide glazing for windows at stairwell and latching hardware for casement windows.

Regarding doors
Door in pocket will be existing reused. Exterior door will be VGF divided light dual glazed. Exterior door hardware will be Schlage standard bore lock and deadbolt. Paint-grade metal top installed on door top to shelter top from water. No mortise locks. As the door is an outswing it will not be warrantied.

Regarding drywall
Bath, hall, kitchen and breakfast nook will be ½-in. drywall with five-point smooth coat or smooth float over existing plaster.

Regarding plaster
Plaster will be patched at foyer side of new pocket door and tied in.

Regarding cabinets
Cabinets will be by Strictly Custom in accordance with the attached specifications and contract. Existing cabinet and top in laundry will be reinstalled.

Regarding interior finish carpentry
Pocket door on rear hallway side and bath door will be cased with 1x3 pine flat stock. (Note that it is not feasible to do plaster returns here).

Regarding countertops
Laminate countertops including nosing extension for tile will be built and installed by Lon Williams. Laminate will be solid color with matte finish Wilson Art or Pionite

Regarding tile
Owners will select tile. It will be purchased by the tile setter and charged to the owners at his cost plus 15%.

Regarding stucco
Two coats by Moses Brown, Kensington Plaster, to match existing work.

Regarding deck
Frame will be pressure treated. Decking will be 2x6 fascia grade.

Lump-Sum Payment Schedule

Client name: Rick Smithfield
Kitchen and bedroom addition,
2025 MacArthur Avenue, Oakland, CA

	Amount	Holdback
Deposit at contract signing	$1,000	
After site excavation and delivery of framing lumber	$10,643	$550
After foundation form	$6,000	$300
After foundation pour	$4,000	$200
After frame of main floor	$7,000	$350
After slabs	$3,000	$150
After rough plumbing	$3,000	$150
After rough electrical	$3,000	$150
After complete frame	$6,000	$300
After doors, skylights and windows in	$4,000	$200
After roofing	$3,000	$150
After exterior siding and exterior trim	$6,000	$300
After substantial completion	Holdback	200% x punch-list value
After punch list	Balance	

Cost of change orders is payable upon completion of change orders.

A builder's contract can specify a series of fixed payments to be made during the course of a job.

completed. Under the latter option, at regular intervals you apply to the clients, or their bank or architect, for payment.

To protect yourself with both lump-sum and percentage-of-completion schedules, you should require in your agreement that the clients pay promptly, within five days for example, of receiving application for payment. You should also collect frequently—weekly is best. Don't let your client get too far out in front. (One builder signed an AIA cost-plus-fee contract allowing only monthly payments. At the end of the project, with $38,000 due, the client said, in essence, "I'm an attorney; come and get your money if you can.")

For added protection with some clients, you may want to require that they set up a project account from which withdrawals can be made only with a two-party voucher carrying both your signature and theirs. Even with such an account, clients can still refuse to sign for project payment. But at least you know that they have the money, where it is and that they can't siphon it off for other purposes.

For the clients' protection, percentage-of-completion arrangements may provide for the retention of a percentage—often 10% of

Percentage of Completion Application

Payment Request

Borrower _DRAYMAN, LOUIS_ Disbursement no. _3_ Date _3/14/91_
Location _432 BOYNTON/BERKELEY_ Branch _S.F. REAL ESTATE OFFICE_
Project _BED & BATH ADDITION/REMODEL_
Tract no. _____ Lots
Exhibit "B" to Building Loan Agreement, Dated _10/19/92_

Item No.	Item	Estimated Construction	Previously Disbursed	This Request	Total Disbursed	Balance Undisbursed
1	DEMOLITION	6,012	3,900	2,112	6,012	0
2	FOUNDATION	9,014	0	9,014	9,014	0
3	FRAME	7,206	1,000	2,600	3,600	3,606
4	FURNACE	2,986	0	0	0	2,986
5	SHEET METAL	2,400	0	0	0	2,400
6	PLUMBING	4,672	600	1,171	1,771	2,901
7	ELECTRICAL	1,186	0	0	0	1,186
8	ROOFING	1,600	0	0	0	1,600
9	DOORS & WINDOWS	2,413	0	0	0	2,413
10	STUCCO	3,331	0	0	0	3,331
11	DRYWALL	1,450	0	0	0	1,450
12	CABINETS	2,010	0	0	0	2,010
13	FINISH CARPENTRY	3,011	0	0	0	3,011
14	TILE	3,697	0	0	0	3,697
15	SUPERVISION	3,000	600	300	900	2,100
16	OVERHEAD & PROFIT	11,311	1,000	1,000	2,000	9,311

Under the percentage-of-completion method of payment, the builder periodically submits an application for payment in keeping with the amount of work that has been completed under each trade.

each payment until the project is complete. Such a high retention can be tough on you, especially when you are low on working capital. The retention can equal a good part of your profit and overhead for a project. If your clients do require a retention, you should provide balancing protections for yourself. Possibilities include:

- Limiting the retention to 5% or less.
- Collection of interest on the retention, preferably at a rate equal to that for any other personal line of credit, and at least a few points above passbook rates.
- Provision for full payment, including retention, for all work completed if the project is stopped through no fault of yours.
- Payment of retention at substantial completion, rather than after the punch list is complete. You can collect most of the retention at substantial completion, and leave the clients with an amount—such as 200% of the value of the punch list—sufficient to reassure them you have a financial incentive to finish the job.

Dual-Signature Voucher

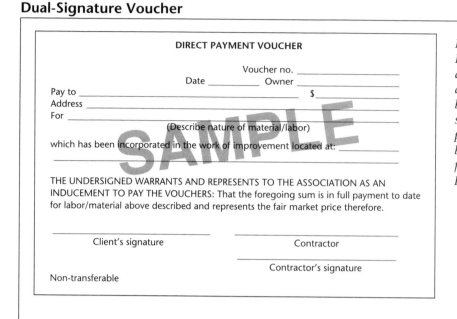

For their projects, Federal Building Company has the clients deposit the funds for the project in a passbook account at Federal's bank. At the completion of each stage of the project, Federal presents the signed voucher to the bank to have funds transferred from the project account to Federal's.

For the sake of both builder and clients, in T&M contracts the distinction between items to be charged as construction costs and those to be covered by the fee for overhead and profit must be clearly spelled out. Otherwise, conflict over payment can easily develop during construction. Items of traditional concern are shown on the facing page. For my own T&M projects, I emphasize to owners that some adjustment and correction of minor errors are a normal construction costs. I show them the lines on my estimating checklist, namely "completion" in both my rough and finish divisions, where I provide 1% to 2% of labor and material costs for the occasional miscut board or door initially hung out of plumb. Though it took me years to realize the fact, I explain, no construction project, any more than other human endeavors, is ever executed perfectly. The occasional human misstep is anticipated as a cost in a lump-sum bid and, by parallel, it is charged as a cost when it occurs under the T&M arrangement. Some owners balk at the notion of paying for errors, and quite understandably. But frankly, they are not good candidates for T&M contracts; under the pressure of construction, they drag you into disputes at the level of "Did your carpenter bend my nails today, and how much credit do I get for the time they spent pulling them?" You are better off working with them under a lump-sum contract.

An agreement should incorporate a rescission notice, allowing the clients time to change their minds and to void the contract after signing it. By the same token, you should allow clients only a limited

Charges Under a Cost Plus a Fixed Fee Contract

In T&M contracts, specify which items are to be paid for by the builder and which by the owner. Shown below are some items that are frequently of concern.

Costs to be reimbursed

Any permits or fees paid for by the contractor in connection with the project.

All costs necessitated by rules, regulations, building codes, taxes, labor rates and government action.

Design and engineering costs if the project is design-build.

Any legal costs incurred by the contractor on behalf of the owner.

Costs of sanitation, job-site phone calls for the project and utilities, if the owner is not paying for such services directly.

Wages of workers, including any project manager, employed by the contractor for the project, whether at the project site or away from the project, for such tasks as cabinet fabrication.

All labor burden, including FICA, FUTA, SDI, state unemployment, worker's-compensation insurance, liability insurance and other costs based upon wages.

Any agreed-upon travel expenses for workers.

An agreed-upon hourly rate for the contractor's supervision of the project, including travel time to and from the project.

A percentage of wages for supplies and services used by the workers but not incorporated directly into the project.

Material, including delivery and tax, incorporated into the project.

Equipment and temporary facilities, including delivery, tax and insurance, rented by contractor for use on the project.

Payments to subcontractors who work on the project.

Cost of bonds directly attributable to the project, unless they are covered by the contractor's fee.

Cost of all cleanup and debris removal.

Cost of action taken to prevent damage, injury or loss in case of an emergency affecting the safety of persons or property.

Losses, expenses or damages to the extent not compensated by insurance or otherwise (including settlement made with the written approval of the Owner), and the cost of correction of work, provided the cost does not exceed the guaranteed maximum price, if any.

Costs not reimbursed

Salaries or other expenses of contractor's personnel stationed at the contractor's office other than a site office.

Expenses of the contractor's office other than the site office.

Overhead and general expenses, except those specified as reimbursable.

The contractor's capital expenses.

Pay of overtime, unless specifically approved by owner or necessitated by an emergency.

Costs that would cause the guaranteed maximum price, if any, to be exceeded.

Notice of Right of Rescission

NOTICE OF RIGHT OF RESCISSION

You have entered into a transaction on (date)_____ which may result in a lien, mortgage or other security interest on your home. You have a legal right under Federal law to cancel this transaction, if you desire to do so, without penalty or obligation, within three business days from the above date or any later date on which all material disclosures required under the Truth in Lending Act have been given to you. If you so cancel the transaction, any lien, mortgage or other security interest on your home arising from this transaction is automatically void. You are also entitled to receive a refund of any down payment or other consideration if you cancel. If you decide to cancel this transaction, you may do so by notifying

David Gerstel
268 Coventry Rd.
Kensington, CA 94707
(415) 524-1039

by mail or telegram sent not later than midnight of_____. You may also use any other form of written notice identifying the transaction, if it is delivered to the above address not later than that time. This notice may be used for that purpose by dating and signing below.

I hereby cancel this transaction.
Date_____Owner's signature_____
Receipt is herewith acknowledged of the foregoing NOTICE, the undersigned OWNER having received copies thereof, this the_____ day of_____, 19_____.
Owner's
signature_____

Effect of Rescission:
When owners exercise their right to rescind, they are not liable for any finance or other charge, and any security interest becomes void upon such rescission. Within 20 days after receipt of a notice of rescission, the creditor shall return to the customer any money or property given as earnest money, down payment or otherwise, and shall take any action necessary or appropriate to reflect the termination of any security interest created under the transaction. If the creditor has delivered any property to the customer, the customer may retain possession of it. Upon the performance of the creditor's obligations under this section, the customer shall tender the property to the creditor, except that if return of the property in kind would be impracticable or inequitable, the customer shall tender its reasonable value. Tender shall be made at the location of the property or at the residence of the customer, at the option of the customer. If the creditor does not take possession of the property within 20 days after tender by the customer, ownership of the property vests in the customer without obligation on his/her part to pay for it.

Clients should be given a few days to rescind a contract. Check with your attorney for the exact legal requirements and proper form for your state.

number of days to sign the contract before it becomes void. If you offer clients a contract that is valid for an indefinite period, you run the risk of hearing from them long after you have forgotten about your offer. Now, they happily inform you, they are ready to build, for the price included in the contract—of course—which has meanwhile fallen far behind inflation.

Finally, an agreement must include room for full signatures. In addition to signing the agreement, you and your clients should initial each document it incorporates by reference. You should provide at least two copies of the contract documents, one for the client and one for yourself. On architect- or government-administered projects, you may also need to provide a copy to the architect or government office.

CONDITIONS

Broadly speaking, the issues covered by the conditions—alternatively called "general conditions" or "stipulations"—of a construction contract divide into three categories: the owners' rights and obligations, the builder's rights and obligations, and mutual or shared obligations. For reasons explained in the previous chapter, you must tailor your conditions to suit your particular needs, drawing as you can from standard contracts. In this chapter we will discuss some of the critical issues deserving special attention, which will, I hope, soften your landing in the dense legal language of the actual contracts you will need to review.

Among the owner's rights and obligations you should consider for your contract conditions are:

- Provision of legal descriptions of the site, surveys, soil reports and plans of underground utilities, including sprinkler systems. The owners should be held responsible for any breakage during construction should they fail to provide correct locations of utilities.
- Providing evidence of clear title to the property. How would you like to learn that your client does not own the land under the foundation you just poured?
- Paying for all permits and any other assessments or fees required for construction, unless otherwise stated in the contract documents.
- Selecting all fixtures, cabinets and like items prior to the beginning of the project. If you let owners slide on selections until work begins, costly bottlenecks can develop.
- Performing or subcontracting work rather than going through you. Although you may want to grant this right to your clients, you should also consider sharply limiting it (pp. 90-91). In my contracts, I provide that owners can directly handle work, either personally or through subcontractors, only before or after my portion of the project. Whenever I ease up on this requirement, owners cause me expensive delays.
- Ensuring free access to and use of the site. For remodels, your conditions should require owners to move personal possessions out of all areas to be worked in, including those you need for staging and storage. If you move your clients' goods, you can be held responsible for breakage, some of which may actually have occurred earlier.
- Keeping children and pets away from the work area.
- Wearing proper safety gear when entering the work area.
- Making required decisions promptly prior to and during construction and compensating you for any delays caused by failure to do so.

Stipulations

Conditions from different contracts use different structures. Gerstel's conditions, called "stipulations,"—of which samples appear below—are simply organized as a series of paragraphs. Do not use these conditions directly without consulting a lawyer.

Arbitration

Any controversy arising out of the construction of the project referred to in this contract, or regarding the interpretation of this contract or any subcontract, is subject to arbitration. The owner, to the contractor, and all subcontractors, and sub-subcontractors are bound, each to the other by this arbitration clause, provided such party has signed this contract or has signed another contract which incorporated this contract by reference, or signs another agreement to be bound by this arbitration clause. Arbitration shall be had in accordance with the American Arbitration Association procedures relating to the construction industry in effect at the time the arbitration is initiated. Should any party refuse or neglect to appear or participate in arbitration proceedings, the arbitrator is empowered to decide the controversy in accordance with whatever evidence is presented. A party shall not be deemed to have waived the right to demand arbitration under this contract by filing suit, provided the party files suit solely in order to obtain the benefit of some provisional remedy, such as attachment or injunctive relief. The arbitration panel shall consist of three members. One shall be chosen by the Owner, one by the Contractor, and they in turn shall choose the third, a professional arbitrator. By mutual agreement, in cases involving a disputed sum of less than $10,000.00, the Owner and Contractor may elect instead to use a single arbitrator.

Changes in the work

During the progress of construction the Owner may order extra work without invalidating this agreement. Any extra work must be first authorized by the Owner. The amount for such extra work shall be determined by the contractor and agreed upon by all parties in advance, if possible. Otherwise it will be done on a time-and-materials basis:
Subs at cost plus _____% for overhead and profit;
Material at cost plus _____% for overhead and profit;
Labor at wages plus _____% for labor burden plus _____% of wages and burden for overhead and profit;
The first _____ hours the Contractor spends writing and pricing change orders will not be charged. Additional hours will be charged to the Owner at _____ dollars ($_____.00) an hour.

Requirements of public bodies

Any changes, alterations, additions to, or deletions from the drawings and specifications which may be required by any public body, utility or inspector shall constitute a change in the work and shall be paid for in the same manner as any other change in the work.

- Paying a licensed specialist to remove asbestos or other toxic material discovered before or during the course of the project.
- Making out a punch list (or having the designer make one out) as the project nears substantial completion. Some contract conditions allow a certain number of days after substantial completion for the punch list to be made. But if you require of owners that if practical it be ready at substantial completion, your crew can take care of it before pulling off the project, which will save you much travel and setup time.
- Receiving lien releases from all subcontractors and material suppliers before making final payment. Although owners should have a right to them, lien releases do involve you in a lot of paperwork. For my projects, I offer to provide releases upon request, but for a fee, to cover the cost of the paperwork.

Filled ground or rock

In the event filled ground or rock or any other materials not removable by ordinary hand tools is encountered, the removal of said material shall constitute a change in the work and be paid for in the same manner as any other change in the work.

Electrical work

Unless specifically included, electrical work involves no change to existing service panel other than the addition of circuit breakers or fuse blocks to distribute electric current to new outlets. Changing of point of service, main switch or meter which may be required by an inspector or utility shall be paid for in the same manner as a change in the work. Changes of existing wiring areas undisturbed by the alterations which are the subject of this agreement are not included by this agreement.

Existing out-of-plumb and out-of-level conditions

Unless otherwise agreed, the Contractor is not responsible for correcting existing out-of-plumb and out-of-level conditions in existing structure.

Matching existing finishes

The Contractor calls the attention of the Owner to the limitations of matching finish surfaces. Although the Contractor will make every effort to match existing textures, colors and planes, exact duplication is not promised.

Conduits, pipes, ducts

Unless specifically indicated, the agreed price does not include re-routing of vents, pipes, ducts, or wiring conduits which may be discovered in removal of walls or in the cutting of openings in walls.

Plumbing

Unless specifically included, plumbing, existing gas, waste and water lines are not to be changed or cleaned.

Termites and rot

The Contractor shall not be obligated to perform any work or correct damage caused by termites or rot, unless otherwise specified herein.

Among the most important prerogatives of owners, and one that impinges sharply on your prospects for prosperity as a general contractor, is the owners' right to order changes in the work. Correct handling of change orders is so important to your survival that I have devoted the whole next chapter to the subject; please read it. But for the sake of our discussion of contract conditions, it is necessary only to note a few basics. First, your conditions should clearly state that owners do have discretion to add to, delete from or modify the project. Second, the conditions must inform the owners that certain types of unwelcome additional work can crop up during a project, especially a remodel, and that they will bear financial responsibility for such work. In my experience, clients frequently do not understand that. During negotiations, they reveal an assumption that hidden problems, such as rot in a wall, will be corrected during the project without extra charge. To avoid disputes, your conditions must disabuse clients of that notion and clearly describe items that will require extra charge. Important areas you may wish to cover are:

- Subsurface soil conditions requiring excavation by means other than ordinary pick, shovel and power spade, unless heavier equipment is specified.
- Extra demolition required by wire concealed in plaster; wallpaper that does not peel away but requires application of a steamer or chemicals for removal; linoleum that does not readily peel off but must be scraped or sanded off.
- Substandard framing in existing structures, including out-of-plumb walls or out-of-level floors and ceilings.
- Damage by rot or insect infestation.
- Hazardous and/or substandard electrical, plumbing or mechanical work.
- Rerouting of pipes, conduits, wiring and ducts hidden in walls.
- Damage from extreme weather conditions or other acts of nature, such as earthquake, despite your good-faith efforts to provide normal protection.
- Requirements by building inspectors for work not called out in the plans or specs.
- Failure of existing work, despite your good-faith efforts to minimize damage, such as plaster cracking or drywall nails popping in adjacent rooms, or blockage of pipes by loosened rust from within the pipes.
- Minor damage, such as scratches and dings, despite your good-faith efforts to provide protection to existing finish surfaces in areas adjacent to or necessarily used during construction.

These final items relating to failure and damage are especially valuable on projects involving existing structures. As one builder says, "Remodeling is an unnatural act," and there's bound to be damage. Without such conditions, an owner (or yourself) may have a hard time knowing how far your responsibility extends. Perhaps, to remodel a kitchen, your crew had to walk back and forth for two months through the adjacent dining room and accidentally (inevitably) scraped the wall a time or two. You need to make clear that while you will touch up the damaged area, you are not responsible for repainting the whole room to match.

Finally, in connection with change orders, your conditions should vest you with authority for determining the cost. Avoid agreeing (as AIA documents ask you to do) that the value shall be determined by the owner or architect as well as yourself. Since their sense of construction costs is often low, you do not want an owner or architect to have authority over your charges. My conditions specify that, if possible, I will give the owner a fixed price for additional work (the great majority of change orders). When I cannot project a fixed price, or when my price is not satisfactory to the owner, the conditions specify that the change order is to be executed on a cost-plus-fee basis. For deletions, you may want your conditions to specify that up to a moderate amount, such as $300, no markup will be credited back. Absorbing deletions into the planned sequence of work costs you. So if you credit back your markup as well as costs on a small deletion, you are likely to lose money.

Your conditions should also require that you will be paid for writing and pricing change orders. On some projects, especially those for which you have received vague, incomplete or inaccurate plans, the work can consume many hours. For the sake of good client relations, however, you may want to give away a few hours of change-order writing. I allow 5 to 15 free hours, depending on the size of the project.

Beyond change orders, your contract conditions should cover a number of other builder's rights and obligations, including:

- Construction of the project in a professional manner in accordance with the plans and specs. The AIA and AGC conditions stipulate also that the builder is responsible for work "reasonably inferable" from the plans and specs. Start-up builders especially may feel that the phrase is unfairly burdensome. (I was sharply taken aback when I first saw it, thinking it gave architects license to demand construction they'd left out of their plans.) But in fact, you will probably find you can work under the requirement, as long as the architect is competent and fair-minded.
- Compensation for design responsibility acquired during a project. If a designer fails to provide correct drawings, you should charge for making the design work when that responsibility falls in your lap during construction. A few days before I wrote this paragraph, for example, one of my leads and I completed the finishing touches on the change order for a custom-window schedule. Virtually every detail from rough opening size through hardware and glazing had been wrong or overlooked in the schedule provided by the architect. When we pushed him, it became apparent he lacked the skill to produce a correct schedule, so we made it up and charged for the work.
- Providing to the owner the names of subcontractors to be used for the project. The owner is thereby protected from your shopping for inferior, cheaper subs once construction begins.
- Determining the organization of the work and selecting techniques and methods. Architects and owners can tell you what to build, but not how to build it.

Delay Clause

Well-organized builders usually get their projects completed on schedule. But sometimes you are delayed for reasons beyond your control. Consult an attorney before adapting a delay clause such as this to your contract.

Delay in starting or completion
The Contractor agrees to pursue work diligently through to completion. Reasonable allowance shall be added to the agreed time for starting and/or completion for the time during which the Contractor is delayed in said work due to inability of the Contractor to obtain a building permit; failure of the Owner to obtain variances, land use permits, easements and utility assessments; or due to acts of neglect or negligence by the Owner; or due to acts of God which the Contractor could not reasonably have foreseen and provided against; or due to stormy or inclement weather which delays the work; or due to strikes, boycotts or like obstructive action by employee or labor organizations; or due to extra work ordered by the Owner; or due to the acts of a public enemy, riots or civil commotion; or due to the inability to secure materials through regular recognized channels which the Contractor could not have reasonably foreseen and provided against; or due to delays in delivery of material or performance of work by materialmen or subcontractors selected for the project which the Contractor could not have reasonably foreseen and provided against; or due to the imposition of government priorities on allocations of materials; or due to the failure of the Owner to make payments when due; or due to inspections of changes ordered by the Owner or by authorized inspectors, or due to the presence of buried or concealed pipes, conduits, water or other utility lines or events not known to the Contractor at the time of this agreement; or due to any additional work due to soil conditions, rocks or fill land; or due to any other cause beyond control of the Contractor and which the Contractor could not reasonably have foreseen and provided against.

- Installing protection for finish surfaces and landscaping in areas to be worked in or used for access during the project.
- Providing for the safety of workers and other people at the job site.
- Making a good-faith effort to match existing textures, colors and other finishes (though not guaranteeing exact duplication).
- Removing debris promptly, cleaning the site daily and leaving it broom clean at the end of the project unless greater cleaning is specified. For my projects, I suggest the owner contract separately with professional cleaners at the end of the project rather than contracting this work through me. They save my markup. I avoid problems. I'm in the building business, not cleaning.
- Nonresponsibility for defects or failures in items the owner provides. (I strongly discourage clients from providing fixtures or other materials for the project, pointing out that down the line they may find themselves caught between a manufacturer and the installer, who each blame the other for a problem.)
- Keeping the project on schedule. For your own sake as well as the owners', you should keep a project moving steadily toward completion. But at the same time, there are many possibilities for delay beyond your control, and your conditions should specify them.
- Providing properly skilled workers for the project and maintaining reasonable discipline among them. Your tilesetter cannot put the make on the clients' daughter. Your drywallers cannot take the lib-

erty of knocking off a couple of six-packs and littering the site with their empties just because they are doing you the favor of working on a Saturday. Your crew can't play rock music at ear-shattering levels to relieve the tedium of demolition. In short, no sex, no drugs, no rock 'n roll. Sorry.

With respect to some issues, contract conditions can provide that both owners and builder have rights and obligations. Both have an obligation to notify one another and the architect if either discovers a defect in the plans. Both should provide certificates of insurance. The builder provides them for worker's-compensation and liability, the owner for course-of-construction insurance, also called "builder's risk," which covers damage to the work by fire, vandalism or other causes while the project is underway.

Either builder or owner can terminate the contract, but only with good cause. You can terminate if the owner fails to make payment when due. The owner can terminate if you persistently fail to build in accordance with the plans, specs and contract conditions, or go bankrupt. In addition, the owners may terminate because of altered circumstances in their lives, such as loss of employment. But if they do, you should then receive compensation for all your costs to that point, including demobilization of the project and also a commensurate portion of your markup.

Because irresolvable conflicts can arise between builder and client during a project, you may want your conditions to require that serious disputes be settled by arbitration. Recent reports suggest arbitration is rapidly losing the advantage it once held over conventional court proceedings. But in general, builders and their attorneys still seem to feel that it can be cheaper and quicker. Some think, also, that given the public hostility toward contractors, you have a better chance with an arbitrator than with a jury. That seems especially likely if your conditions require that the arbitration panel be made up at least in part of contractors. My conditions specify that for disputes involving more than $10,000, the arbitration panel shall consist of three people: a contractor chosen by myself, a person chosen by the owner and a professional arbitrator selected by the first two. For smaller disputes, the conditions specify selection of an arbitrator by mutual agreement.

You have a good chance of avoiding the need for arbitration if you have a complete and fair set of conditions coupled with a proper agreement. When you are ready to contract with a client, don't merely fill out the document and send it over in the mail. Set up an appointment and review the contract with the client. Point out the most important and most potentially controversial conditions, and explain the reasons behind them. Ask your clients to review the entire contract and to raise any questions or doubts. Be prepared to negotiate any sticking points, as long as the contract remains fair and covers all important bases. Remember, a construction project is not won when your bid, whether competitive or price planned, has been accepted. The project is yours only when both parties are happy with the contract and it has been signed.

Change Orders

If your estimate was your first chance to leap to financial catastrophe on each project, then when you come to change orders, you've reached a yawning stairwell down to disaster. My personal lesson on change orders came about five years into my contracting career. I had won my largest job yet, with what I thought an ample bid. When extra after extra cropped up during the rough work, I did the work without presenting the owners with a change order. I figured I was charging these nice people enough (I was even a little guilty about my anticipated profits), and so much money was flowing my way that I though it could never run out. By the time I saw my mistake—as bills began to come in from subcontractors and suppliers of expensive finish materials—it was too late. I completed the job with only half the earnings I had expected.

Years later, when I began research for this book and compared experiences with builders who had fallen down the change-order well, I realized my lesson had come relatively cheaply. When Laura Lloyd, for example, looked back on the change-order disaster that occurred during her first major project, she reflected that she "had started down the slide even before the job started." Hungry for work and an exciting project, Lloyd had contracted for a large, complex remodel with a couple who "wanted everything now, but as cheap as they could get it." They cajoled a low price from her after rejecting higher bids from several more seasoned contractors. As the job progressed, the clients repeatedly asked Lloyd for additional work, including a master bathroom addition for which plans were drawn up by their architect. Lloyd accepted their requests, but never wrote a change order. "I didn't imagine they could contest extra charges for work not shown on the architectural drawings," she said later. But the clients did contest the charges, and in fact refused to pay Lloyd's final bill for $60,000. Then they sued her for fraud and sought to have her license revoked. In arbitration, she was awarded $33,000, just enough to pay her legal fees.

Even well-established operations are not immune to change-order disasters. For example, the Waldorf Company, a firm known for the completeness of its business systems, won a contract for the renovation of a civic club. Waldorf himself, a contractor with 35 years of experience, kept an eye on the project and carefully instructed his lead, a new employee, to make sure to get all orders for changes from the original plans in writing. But he did not follow up with the lead to make sure the "writing" was a formal change-order document signed by an authorized representative of the club. The lead, trying to do his job just as instructed, conscientiously kept notes describing all additional work he performed. At the end of the project, Waldorf noted that there was a substantial difference in scope between the finished project produced by his company and the original plans. He figured he was in line for a nice extra profit. Then he discovered that he had no written proof that the club had authorized any of the extra work. After a long wrangle as to just what was extra and what not, and how much of the additional work had been authorized, Waldorf had to settle for half of the $41,000 he felt he was due.

The saga of change-order catastrophe continues onward to the kinds of firms that handle major urban projects. My absolute favorite change-order story comes from Washington D.C. A general contractor named Kirten/James built a large center for a well-known clothing retailer. At the end of the project, the contractor sued the retailer for $3.9 million for refusing to pay for changes from the original plans. Incredibly, Kirten/James's attorney justified his client's claim by arguing that the retailer "ordered changes throughout the course of the project, frequently on a verbal basis or written on napkins or walls." Could the attorney have dreamed up a more graphic declaration of his client's incompetence? A builder who is on top of a project may, to keep work going, accept a verbal change order for a few hundred dollars as long as it is agreed that a written authorization will speedily follow. But napkins? ("Tell us, Mr. Attorney, did your client favor stray napkins from workers' lunches or fresh ones from the hamburger place around the corner? Did your client make carbons directly from the original or go in for napkin photocopying? Or was he or she able to develop a technique for napkin faxing?" Here we have the makings of a construction-industry comic strip.)

Obvious lessons shine through these disaster stories. So why then do builders repeatedly fail to stay on top of change orders? With experienced companies, the problem seems to stem from breakdowns in communication between office and field, poorly developed management systems or sometimes a bout of undue relaxation resulting from sentimental attachment to a client or project. With start-up builders, the source of the problem is often naivete. But it can also be

Change-Order Form

Fold at (>) to fit 771 DU-O-VUE® Envelope

PRODUCT 271-3 [NEBS] Inc. Groton, Mass. 01471. To Order PHONE TOLL FREE 1 • 800-225-6380

DAVID GERSTEL BUILDER
Lic. # 325650
268 Coventry Road
KENSINGTON, CALIFORNIA 94707

(415) 524-1039

CHANGE ORDER

Number

TO

PHONE	DATE
JOB NAME / LOCATION	
JOB NUMBER	JOB PHONE
EXISTING CONTRACT NO.	DATE OF EXISTING CONTRACT

We hereby agree to make the change(s) specified below:

NOTE: This Change Order becomes part of and in conformance with the existing contract.

WE AGREE hereby to make the change(s) specified above at this price ⇨ | $ |
DATE | PREVIOUS CONTRACT AMOUNT | $ |
AUTHORIZED SIGNATURE (CONTRACTOR) | REVISED CONTRACT TOTAL | $ |

ACCEPTED — The above prices and specifications of this Change Order are satisfactory and are hereby accepted. All work to be performed under same terms and conditions as specified in original contract unless otherwise stipulated.

Date of acceptance _____

Signature _____ (OWNER)

A simple mail-order form is acceptable for small projects with few and minor changes. But a professional change-order form — such as the one shown on the facing page does more than show the cost of each individual change order — it records the cumulative cost of all change orders on the project, and it also extends the required date of substantial completion to include the change.

discomfort at presenting clients with the change-order costs. In fact, even after two decades as a licensed contractor, I still experience that unease, especially when presenting sizable charges for the necessary correction of hidden conditions on remodeling projects. Clients groan as you tell them the $1,800 they had hoped to put into new furnishings must go to replace the rotted framing or rusted pipes you discovered after opening the walls. It's not fun being the bearer of bad news. But unless you want to join the ranks of Lloyd, Waldorf and Kirten/James, you must develop and vigilantly enforce a sound change-order procedure. Here are some guidelines:

• Acquire or develop a good change-order form.

• At the bidding and contract-writing stages, make clear to your clients that extra costs will (not may) arise during construction. Insist they keep in reserve contingency funding beyond the contract price to cover hidden conditions and design changes. In my experience, contingency funding typically needs to be set at 5% to

Change Order with Calculations

CHANGE ORDER

David Gerstel/Builder
License #325650
268 Coventry Road
Kensington, CA 94707
(415) 524-1039

To: *DONNA & ROB JENNINGS*   Number _5_

Job name / Location: *BOATHOUSE / 2 LAKE WAY*

We hereby agree to make the following changes in the work:

● *INSTALL DECK IN ACCORDANCE WITH PLANS
BY RICHARD MORRISON, pgs 1-2, 3/11/91*

Cost of this change: *$2,905.00*

Total cost of all changes to date: *$6,320.00*

Change in construction schedule from this change: *5 days*

Total change in construction schedule: *11 days*

Contractor's hours: On this CO & on total CO's: *3 & 7*

Owner's signatures: *Robert J. Jennings* Date: *3/15/91*
Donna M. Jennings Date: *3/15/91*

Contractor's signature: *David Gerstel* Date: *3/15/91*

A fair construction contract provides for the builder to be paid for writing change orders. In practice, Gerstel typically writes 5 to 15 hours worth without charge, as an investment in client good will.

10% of the contract price. For renovation of older, poorly built and run-down structures, 50% and even 100% may be necessary. Upgrades in finishes can push the percentage yet higher.

• Write and get signatures on a change order for any work not included on the plans and specs. On some of your projects, architects or engineers may write the change orders. But typically, on projects handled by small-volume builders, responsibility for writing change orders falls to the builder.

• Treat the pricing of the change order as you do any other estimate, taking special care not to omit hidden items such as difficult access. If the change order is large, use your estimating checklist when pricing it.

• For the sake of maintaining trust during a project, don't nickel and dime your clients on small changes. Write those little extras up at no charge, so clients can see you're not squeezing them at every opportunity. Then both you and your clients will be more comfortable with the larger changes for which you must charge.

- Do not begin work on an extra until the change order has been signed by the client. Make exceptions only for low-cost changes, when you need a quick verbal approval so that your crew can keep moving. Get a written backup as soon as possible.

- Collect for change orders soon after the work is complete. Do not wait until the end of the project, especially for sizable amounts. If you do, you may find that your clients do not have enough to pay you, or that they are so surprised at the size of the bill that they refuse to pay you. Note that in the Lloyd and Waldorf change-order stories, along with not writing change orders, the builders waited until the end of the project to charge for changes. Had they billed earlier and the clients refused to pay, they could have minimized their losses by terminating the project.

Change orders will never make you popular. That's one reason they so often go unwritten; builders fear that by presenting change orders they will incur their clients' dislike. But change orders, in addition to keeping you solvent, maintain the complete communication that is the foundation of successful construction management. Once I was told by a happy client that she liked working with my company because there were no surprises. The cumulative totals on the change orders kept her informed about both the financial status of the project and extensions in schedule. She always knew where she stood. Perhaps to be a successful contractor, you must be more interested in being respected than in being liked.

 # SUBCONTRACTS

If you have developed a comprehensive and fair agreement and conditions and a thorough change-order procedure, you have gone far beyond the typical small-volume builder in the matter of contract documents. You may feel no need to go farther and be content to operate without a separate document for your subcontractors. Instead, you may choose to contract with subs by either verbal agreement and phone quotes or by contract documents that they provide. If that is the path you prefer, you will find some persuasive rationalizations to tread it:

- If you are a well-organized builder who creates job-site conditions that allow subs to work efficiently and who pays them promptly, subs will want your repeat business and, therefore, will fulfill their obligations.

- You may work repeatedly with the same subs. You develop mutual trust and knowledge of one another's methods. On your typical projects, expectations are clear before they are written up in a contract and any details are easily worked out during construction.
- The effort of getting written contracts in place may not be worth the protection gained, given the relatively small size of your contracts with your subs.

On the other hand, you may want to walk the last mile and create a subcontract for at least occasional use. You may want to have one on hand for the first few times you try out a subcontractor. You may also want to be able to pull it out when you contract with even a regular sub for a job that involves an unusually large dollar volume of their work.

Subcontract Short Form

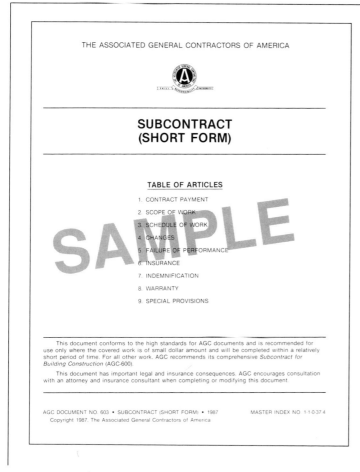

The table of articles of this AGC subcontract can function as a checklist when you draft your own subcontractor contract.

If you do elect to create a subcontract, like your prime contract for owners, it should be balanced and fair, not designed to protect only you. Subcontracts are generally much shorter than prime contracts, and therefore can combine an agreement and conditions into a single document. Among the issues a subcontract should address:

- Identification, including names of contracting parties, location of project, designer's name and date of signing.
- Description of the project and scope of the work. The subs must contract for all work shown on the plans and specs within their trade unless an item is specifically excluded.
- Clarification of responsibility for items that can fall between trades. On most projects, many items could be assigned to either a sub or your crew. For example, some plumbers expect to set and hook up dishwashers; others hook them up only after the general contractor has set them. If responsibility for such an item is left out of everybody's estimate, somebody has to eat it. You don't want to eat it. You also don't want to force it on your sub and sour a working relationship. (See pp. 115-125 for checklists to help eliminate oversights.)
- Permits. Unless otherwise specified, subcontractors should draw their own permits for the project.
- Coordination with your prime contract. The subs perform their work in accordance with the general conditions of the builder's contract with the clients.
- Payment schedules. Normally, subs receive partial payment after completing each phase of their work. For example, my electricians typically get two-thirds of their contract price after rough-in and the balance after finish. However, if the owner retains a percentage of the general contractor's payment until substantial completion, the general contractor retains the same percentage from payments to subs.
- Notification and schedule. The builder gives the subs ample notice of when their work must be performed during a project. Subs are then expected to begin on time with a work force adequate to get the job done on schedule.
- Overtime. Some subcontracts require that, if necessary, subs put their workers on overtime to meet production schedules. But you should consider that such a clause may alienate your subcontractors by placing them at risk for severe losses. Subs work for many general contractors, and sometimes they find themselves called by everybody at once. They cannot meet production targets on all their projects without overtime, but the cost of overtime can put them deep in a hole.

- Failure to perform. If a sub fails to perform on schedule, you have the right to cancel the contract with the sub and to seek another. You need the clause in place for the instances when you cannot wait any longer without bogging down the project.
- Change orders. You can require subs to perform extra work ordered by the clients. Just as you retain the right to price changes in your general conditions, you should extend the right to your subs, and provide that if a fixed price cannot be agreed upon, the work will be performed on a T&M basis.
- Back charges. You deduct from payments due to subs the cost of repairing any damage they do at the project, if they fail to take care of it themselves. Again, invoke the clause sparingly. You do not want to nickel and dime, and thereby alienate, a good sub.
- Failure of materials and products provided by the clients or by the general contractor. Just as you are not responsible for failure of products supplied by the owner, neither are subs responsible for products they do not provide.
- Material and equipment. Unless otherwise specified, subs provide all material and equipment needed for their work.
- Job-site behavior. Same deal as for yourself and your crew. No sex, no drugs, no rock 'n roll.
- Cleanup. Unless otherwise specified, subs are responsible for their own cleanup. (Frequently, it will be more cost effective for you and your crew to handle cleanup than to leave it to the subs.)
- Insurance. Subs carry worker's-compensation and liability insurance and provide the builder, the clients and the designer with certificates upon request. Associated General Contractors suggests that the clients and the builder should have themselves named as "additional insureds" on the subs' policies.

In addition, subcontracts can contain provisions paralleling those in your prime contract on the issues of workmanship, protection of work, inspections, termination and arbitration. To create your own subcontract, you can use a procedure similar to the one suggested for prime contracts (see p. 144). One last time, I advise you to invest in an attorney's review as part of the process.

PART 7

LABOR AND MATERIAL

Hiring and Firing
Crew
Crew Leaders
Pay
The Four-Day Week
**Subcontractors and
    Suppliers**

HIRING AND FIRING

"Good help sure is hard to find," you often hear builders complain. In their version of things, everyone who swings a hammer for a wage is a deadbeat. Meanwhile, however, you frequently see successful contractors fielding able crews. So you can't help wondering if the problem is not so much with inept workers as with inept management. Perhaps the complaining builders just are not hiring right in the first place.

Hiring demands great care. Prepare to search for as long as necessary to find a good candidate. What should you look for? Some builders look for experience and skill. Others are more concerned with character—they figure they can teach people how to measure, cut and nail, but integrity is either there or it's not. In my own hiring, I find that I look for everything at once: dependability, the ability to build accurately and efficiently and to absorb new skills rapidly, ambition and a certain amount of assertiveness. But I also want someone who is friendly and courteous to clients, other crew members, subs, and last but by no means least, to me.

When I hire, though I look for a wide array of strengths, I don't dismiss a candidate because of one or two weaknesses. In fact, I would rather have a person with several considerable strengths and a significant weakness, with four A's and a D+ so to speak, than somebody with five B–'s. It seems to me that people with strengths will have what it takes to do something about their weaknesses once they are pointed out.

When you first hire, though the natural tendency is to look for a low-wage apprentice, you should actually seek someone who can quickly mature into a crew leader. With an employee to whom you can entrust leadership at the job site, you gain the flexibility you need to manage your company properly.

Prospective employees typically come your way by one of three methods: by chance, by reference or in response to advertising. Thus, the man who is now my senior lead walked onto my job site 12 years ago. Several times I have met good laborers and carpenters in bookstores and at the sports bar where I watch NBA games on the big screen. Now, whenever I encounter likely candidates for future employment, I file their names and phone numbers. When it comes time to hire, I have several likely prospects at hand.

To enlarge your pool of prospects beyond those brought by chance, you can seek references—especially from other members of your builder's association (see "A Plan," pp. 13-20). I have found that I can quickly gather several likely candidates for any position in my company with a couple of hours of telephoning to other members of my group.

As contractors become more seasoned, they tend to rely less on chance and references to locate prospective employees and more on ads. But different types of ads can produce very different results. The common want ad, for example, often attracts too many applicants, who you must then tediously sort through. One builder echoes the frustration of others I have talked with when he says that hiring by means of a want ad is like "wading through garbage; it's a last resort." Too many of the respondents are looking for any job, and come nowhere close to the requirements stated in the ad for your job. Not only do apprentices apply for journey-level positions, and vice versa, but people who have never even worked in the trade apply. With ads, as with other aspects of contracting, you should ask around to learn what's effective in your area and what's not. In my town, for example, an ad placed on the bulletin boards of the best building-supply companies will often produce a stream of promising applicants.

Reliable temporary workers at any skill level and leads can be especially difficult to find. In the past, unions were a primary source of temporary labor. If you needed extra hands for a few days, you called the hall and they sent the workers. Today, small-volume builders rarely operate union shops. They have had to look to alternative sources: college dormitories or government employment offices. One builder I know has tapped into a circle of aspiring actors who keep themselves afloat with intermittent construction work.

Sources for Potential New Employees

If you keep a record of prospective employees, when you are ready to hire you may find the person you need in your file. Your next step may be to ask around for recommendations. If you don't find the person you need among the prospects or references, you may need to advertise. But be prepared to interview many people to find a few good prospects.

Prospects file
People who have applied for work in the past
People you have met
People you have heard about

References
From current employees
From friends
From other builders
From subcontractors
From inspectors

Advertising
At lumberyards
At trade schools
In newspapers

Ad for an Apprentice

An ad placed on the bulletin board of your best building-materials supplier can draw promising applicants for a job.

Apprentice Carpentry Job Available
with David Gerstel/Builder

Requirements:
- Midlevel apprentice skills; you must be able to measure, cut and install lumber to layout by an exacting crew leader
- Neat appearance, courteous manner
- A commitment to learning the carpentry trade the right way

Benefits:
- $12/hr. to start
- Legitimate paycheck. No under-the-table or "subcontract" shenanigans. You will be fully covered by disability and worker's-comp insurance.
- Four-day work week: Tuesday through Friday, 9½ hr./day, with 45-min. lunch breaks on your time, and morning and afternoon breaks on ours.
- Clean, safe work sites
- Top-quality workmanship
- Good possibility of long-term employment on one challenging project after another
- Medical plan

For an interview, call (401) 423-1817
and leave a message. Your call will be returned.

Increasingly, however, small-volume builders do not need to rely on such ad-hoc sources. Among the fastest-growing service industries in this country are the temporary labor agencies. In my area, several specializing in construction work have opened in the last few years. With a phone call, you can usually get people of any skill level for the next day. At the best (and most expensive) agency, the workers are screened and tested and their performance is monitored, so that you usually get reasonably capable individuals.

Temp agencies offer an additional attraction, in that they, not you, are responsible for their workers. Therefore, even if a worker is hurt on your job, the injury is covered by the agency's policy and does not effect your worker's compensation rate. When temp workers are finished at your site, they head back to their agency, not to the unemployment office to make a claim against your account. But temp workers are never a replacement for a permanent crew. The difference in commitment—yours to them and, therefore, theirs to you—shows up in performance. Still, temp workers from a well-run

Ad for a Crew Leader

Are You Ready to Run a Job?
Lead needed by David Gerstel/Builder

Requirements:
- A minimum six years full-time employment as a carpenter
- An ability to execute all rough carpentry tasks accurately and efficiently, including layout, form building and framing of walls, floors, roofs and stairs
- Long-term commitment to construction; you intend to make a career of it
- An ability to lead both apprentices and journey-level workers
- A reliable truck and a full complement of manual and power hand tools
- References for work experience and skill level
- Examples of your work available for viewing

Benefits:
- $20/hr. to start
- Merit raises
- Tool allowance
- Health plan
- Profit-sharing after first year
- Legitimate paycheck. No under-the-table or "subcontract" shenanigans. You will be fully covered by disability and worker's-comp insurance.
- Four-day work week: Tuesday through Friday, 9½ hr./day, with 45-min. lunch break on your time, and morning and afternoon breaks on ours.
- Clean, safe work sites
- Top-quality workmanship and good crew morale
- High possibility of long-term employment on one challenging project after another

For an interview, call (401) 423-1817
and leave a message. Your call will be returned.

The best respondent to an ad like this, placed on your supplier's bulletin board, is likely to be a top carpenter from another company who is ready to move up, but can't because all the company's leads are well established.

agency can be enormously helpful. I rely on my agency to beef up my crew for demolition, excavation and any large quantity of simple structural work at the start of a project.

Looking for a capable lead creates far more pressure than hiring for other levels. You know that if you make a mistake and entrust project leadership to the wrong individual, you can pay an enormous price. One builder says the best potential leads are contractors who are in business for themselves only because they were never able to find steady employment with a good builder. They'd much prefer to be running a job site and working with the tools, leaving other management to a supportive and capable boss.

Other possibilities for a lead include carpenters eager to move up, who don't have that opportunity with their current employers. The best candidates may (even should, if you have planned ahead) already be in your company. For my company, whenever I hire at any level, I now look for an individual who has the potential to become a lead in time. Then, when I do need to fill a lead position, there will likely be someone promotable already in the company.

Items to Cover in a Written Application

- Name, address, phone number
- Social-Security number
- Education
- Three previous jobs, with names and phone numbers of supervisors, dates of employment, wage level and types of work done
- Amounts of experience in main divisions of your work, such as demolition, foundations, framing, doors and windows, siding, trim, cabinet installation and subtrades
- Reason for applying
- Long-term ambitions
- Vehicle and insurance company
- Tools
- Vacation expectations
- Acceptablility of overtime and/or weekend work
- Minimum starting pay required and expectations for wages
- Person to contact in case of emergency
- Signature and date

Note: There are a number of subjects you cannot legally ask about, such as weight, physical condition, race and religion, without violating a person's civil liberties. Have your attorney review your written application before you start using it.

Once you have attracted a pool of candidates for a job, you need a coherent procedure for sorting through them. You might use a traditional five-step process: initial phone contact, written application, in-depth interview, reference checks and trial employment.

You will often eliminate many applicants with the initial phone contact. Other callers you will invite to fill out a written application. Take care here. By law, there are many items, such as race and creed, you cannot inquire about without violating the applicant's civil rights. The list of prohibited items seems to expand with the passage of time. If you do decide to use a written application, have it checked by an expert for legality. Typically, written applications ask for education and work experience, but some builders have found they can also use them to evaluate skill. One builder asks applicants to list all the steps involved in several rough and finish operations. Another asks carpenters to label parts of a house frame.

When I hire, I consolidate the first three steps—phone contact, written application and in-depth interview—into one. If a caller seems initially promising, I prolong the conversation, taking notes on the application form as we go. Earlier in my building career, I interviewed by working my way down a list of specific questions. But I found that people responded with stiff attempts at the "right" answer. Now I use a more conversational approach, encouraging the applicant to talk by means of such broad questions as "How about if you tell me what you've been up to and what you're looking for in your construction career?" I avoid listening in dead silence. It intimidates. I frequently say "yes" and "interesting" and "I'd like to hear more about that." I also encourage the applicant to take the lead, to raise personal concerns and to ask questions. What quality of work do we do? Are the projects challenging? Are there opportunities to learn? What about safety practices? What are the experience and skills of other people in the company? The questions give me a sense of the applicant's values and goals. If the focus is entirely on wages and benefits, that also influences my evaluation.

Usually the conversational interview will quite naturally hit most of the specifics on my checklist. Any it misses, I cover directly. To assess trade knowledge, I ask an applicant to describe his or her experience in all phases of construction—in the sequence they normally occur during a project. The response gives me a sense not only of specific skills, but of overall understanding of the construction process. I have found, also, that if applicants describe procedures in the language of the trade, there is a fair chance they've learned it from pros. The person who says "I've done a lot of nailing on walls" is not likely to be a framer. The one who says "Oh yeah, I've stacked, stood, plumbed and lined my share, and rolled quite a few joists, too" may be.

I also try to get a sense of applicants' overall interests and circumstances. I want to know if they have a commitment to construction. Do they look upon it as a career? Or is it merely a way to pay the bills until they decide what they really want to do when they grow up? Have they been able to hold a long-term job? Or have they always bounced around in their work lives? Equally important are applicants' lives outside of work. The law prohibits me from asking certain specific questions here, but during our conversation I'm hoping to hear of consistency in some aspect of life. I'm not in agreement with the famous entrepreneur who says he hires only people who have proved they care about others by staying married. But I do feel that people who cannot bring stability to any aspects of their lives are unlikely to become reliable members of a construction company.

With telephone interviews, I can narrow the field to a few candidates. Checking references is the next step—and a crucial one, as I have learned the hard way. I phone at least three references. I ask the employers to describe the work the applicant has done. I check their description against the applicant's and also ask them to comment on reliability, productivity and communication skills. I keep in mind that some employers are afraid to give a bad reference, since under current law they can be sued for doing so. Hesitation or evasion signals to me that something important may be going unsaid: "Patrick was...uh...okay...a good employee." "So you'd hire him again?" "Huh!...oh...well...we're not hiring right now." If, on the other hand, I get prompt, direct and enthusiastic responses, I think I'm on the right track: "I hated to lose her when she moved out your way. She was fun to work with, a quick learner and she loves carpentry. I'd hire her back in a minute."

After checking references, I arrange to meet the single most promising applicant, usually at the job site. A couple of times a well-recommended applicant who made a good phone impression immediately raised my doubts in our face-to-face meeting. Maybe the vehicle was dirty, or the applicant was unkempt or in woeful physical condition. I do take personal neatness as a precursor to conscientious work. And, for the sake of safety, I do not want to put someone who is badly out of shape on a construction site.

If a candidate continues to make a good impression after a few minutes of conversation at the site, I give a skills test. For apprentices or journey-level workers, I walk the candidate around the site asking him or her to name different items of material and to describe different elements of the project. I also ask for performance of a few tasks,

Procedures for Firing to Minimize Unemployment-Insurance Costs

If you allow workers to file invalid claims, they drive up your unemployment-insurance rates. To prevent invalid claims:

• Document any violation of company rules or standards (such as tardiness to work or neglect of safety practices) prior to termination.

• Issue a written warning when rules are ignored. Be specific about the consequences. Have the employee sign and date the warning.

• Conduct an exit interview at termination. Ask the crew leader to attend. Explain why you are terminating the employee. Let the employee respond. Help the employee develop ideas for a new job.

• Notify the local unemployment office of the employee's termination.

• If the employee makes a claim for unemployment after termination for a good cause, fight the claim.

such as stacking full sheets of plywood, cutting a list of blocks to length or making a miter joint. With leads, I'm interested in a demonstration of layout techniques (for example, of a stair stringer and bottom plate) and of list-making, to assess how the candidate would organize a workday.

Sometimes it is soon clear I've been optimistic about a candidate's skill. He or she simply doesn't have the moves. For example, when one applicant for a mid-level apprenticeship struggled to get a sheet of ½-in. plywood off the ground, then brought me a 12-ft. 2x8 instead of the 8-in. 2x12 block I had requested, I realized I had to go on to the second choice on my list. On the other hand, when another candidate rattled off the names of the hardware items in our extensive on-site storage system, I figured she must really be the quick learner her references had claimed.

As a last step, I ask applicants to read our company policy, and to tell me if they find it acceptable. If so, I will likely offer them a tryout—not a permanent job. In my experience, the tryout really has three phases. After a few days, you or your lead know if you have severely overestimated a worker's reliability or skills. After a few weeks, you can often see whether a new employee is merely marginal. After about three months, any honeymoon is over, and some individuals who have started off strong will begin to flag and take liberties. Others will still be pushing hard, and will have established themselves as productive members of your company.

Sometimes you must fire an employee. You've made a mistake and hired the wrong person, or a once capable worker has gone stale. There are two basic reasons for firing employees: for violations of basic company policy (repeated lateness to work or refusal to wear safety gear, for example) and for serious and persistent slippage in performance. The employee is not maintaining focus, but is making constant mistakes, dragging at the work. Perhaps he or she is distracting others on the crew with constant chatter or grinding down their spirits, and yours, with harsh and insulting remarks.

When you do fire employees (as opposed to laying them off for lack of work), they are not eligible for unemployment benefits. To protect yourself against illegitimate claims, follow the procedures outlined at left. In addition, for the morale of both the employee you are firing and the crew that stays on, handle the firing with consideration. Do not fire anyone unilaterally. As both a reality check on your own judgment and a gesture of respect, ask the lead before making a final decision about anyone on his or her crew. Once I was deterred from firing an apprentice by the lead, who said he would straighten the kid out. He did, and the apprentice has matured into a reliable and good carpenter.

Warning Letter

A warning letter can help you ward off an invalid unemployment insurance claim. It can also turn around an employee. After signing Gerstel's letter, young Both ceased coming to work late and matured into a great apprentice.

```
4/7/88
From: David Gerstel, General Contractor
License #325 650
To: John Both
Regarding: Repeated lateness — Pretermination warning

Dear John,
Since I hired you in December of 1987, you have
repeatedly been late to work. Several times I have
talked to you about your tardiness and explained
that it is a problem you must correct if you want to
continue working for me.

Nevertheless, on Tuesday, April 5, 1988, you did not
show up at the job site on time. John Jenkins, your
lead, called your home and found that you were still
asleep. You made it to work only a few minutes
before lunch.

Therefore, please be advised that if you do not
correct your lateness problem, you will be
terminated.

Please sign this letter and return it to me. Your
signature will be taken as an indication that you
1)agree that the statements in the letter are
accurate, 2) understand the letter and 3)would like
to continue working in my company and are making a
commitment to correct your problem.

Good luck.


David Gerstel

John Both  signature_____Date_____
cc: John Jenkins
```

With some employees, especially people who have been with you for a long time, you may wish to precede firing with a warning letter, or perhaps a candid talk describing exactly what changes are necessary to resecure their jobs. When you finally reach the point of firing, it's best to make it quick and clean. Do it at the end of a workday. Have a final paycheck ready and, for a long-term employee, a week or two of severance pay if you can afford it. Some builders advocate "no discussion" at a firing. But I think it's better to offer departing employees an explanation and let them decide if they want to hear it. Give them a chance to fire any parting shots. Then you can agree to disagree, shake hands and separate on relatively cordial terms, thereby strengthening rather than weakening your company.

Everyone who has worked with your company—all employees, as well as clients and other building professionals—are part of your network. Every employee who leaves bitter will damage it.

When the firing is done, explain it to the rest of the crew. Make clear that their jobs are not in jeopardy. I recall from my own days as a carpenter that when someone was fired, the rest of us tensed up, wondering whose head would be the next to roll. Morale and productivity declined. A phrase such as "This guy was just not doing the job to our standards" will reassure your crew.

CREW

Hiring the right people is just the first step. You must then organize them into crews of the right size, with the proper mix of skills, and give them the support they need to work efficiently and safely. Optimal crew size is a matter of debate among builders. Some think three people is ideal. Others prefer two, on the premise that the lead can then really produce rather than supervise. Walt Stoeppelwerth, a publisher of remodeling manuals, reports that studies show the most effective remodeling crew is one person.

It seems to me that with crew size, as with other issues for which we often hear formula solutions, there are no across-the-board answers. Different projects and circumstances demand different crews. I do feel, however, that small-volume builders often maintain crews that are inefficiently large. I made the mistake myself until one of my three-person crews shrank to two, and I saw that daily production dropped much less than I had expected, by perhaps 10% instead of 33%. From that experience, as well as from talking with other builders, I have come to this rough rule of thumb regarding crew size: Up to three people, the crew for a project should be kept as small as consistent with reasonable progress. In other words, one is generally better than two, and two better than three. Once you exceed three people, field as large a crew as practical—with the limiting factors being the availability of good people and the number that can effectively fit onto the site. Here is my reasoning:

When you add a person to a crew, you do not gain a full person's worth of productivity. As crews grow larger, increasing amounts of time can be lost to the increasingly complex interactions between members. Moreover, the new person will require supervision by the lead and thereby lower the lead's production. (That is one reason why my crew that shrank from three to two lost little productivity. The lead was free to do more hands-on work.)

A good lead can do hands-on work while still supervising and laying out for a couple of workers, especially if they are skilled or the work is repetitive. But with more than two people to supervise, the lead becomes so distracted that his or her hands-on production will markedly diminish. At that point, the lead may as well go over to full-time supervision, and take on as many workers as practical.

As with crew size, small-volume builders differ in their opinions of ideal crew mix. More seasoned builders often prefer to work largely with journey-level people, and for good reason (as the chart below shows). Start-up builders often are not aware of the real cost of apprentices, and attracted by the relatively low wages, tend to create bottom-heavy crews. But even on simple work, the productivity of the apprentice will likely not compare to that of the experienced person. And, as work grows more complex, especially in remodeling, where the problems are so varied, the apprentice is likely to fall far behind.

Of course, rules of thumb about crew size and crew mix are no more than starting points for your thinking. Your crew size and mix will be shaped in response to actual conditions. On some projects, size and mix may need to change as work proceeds. For example, for my projects I use permanent two-person crews—a lead and a talented apprentice who works hard, learns new tasks quickly and seems to have the potential to lead a crew someday. Some jobs the two can handle entirely on their own. For others, we get them help—workers from our temporary agency during the rough stages, and independent licensed finish carpenters later in the project.

The True Cost of an Apprentice (Per Hour)

	Apprentice	Journey level	Crew leader
Wage	$10.00	$16.00	$20.00
50% labor burden	$5.00	$8.00	$10.00
Total hourly labor rate	$15.00	$24.00	$30.00
Supervision cost	$5.00 (10 min. of lead time/hr.)	$1.50 (½ hr. of lead time/day)	$0.00
Tool cost	$3.00	$2.00	$1.00
Cost of correcting errors	$2.50 (10 min./hr.)	$2.00 (5 min./hr.)	$1.00
Real cost	$25.50	$29.50	$32.00

If you look at wages alone, apprentices may appear to cost only half as much an hour as a lead. But if you add in labor burden and the cost of supervising them, the gap narrows. When you figure that they are likely to make more mistakes, to make more use of company tools and to be far less productive at most tasks than a lead, their true cost appears still higher.

Crew size and mix will also be influenced by the workers available to you. Obviously, a crew of less than optimum size but of good workers is better than a mathematically correct number of goof-offs. Far and away most important, crew size and mix must take into account employee morale. You might feel, for example, that on a particular project you would ideally bring on board a journey-level carpenter for the framing. If you did, however, your reliable and loyal apprentice would go without work. So the apprentice gets the job.

Take care of your people, and they will take care of you. This axiom and its consequences are understood by successful entrepreneurs in all industries. It is particularly important in construction, where workers can so easily change employment. If you don't treat them well, they can pick up their tools and move down the road to a job that won't be any worse and might be better. Or they can simply go into business for themselves. Without a steady crew, you get into the kind of trouble Bob Syvanen describes in his admirably honest book, *What It's Like to Build a House* (see Resources, p. 223). Syvanen is working with a pickup crew, and every few pages it trips him up. His guys don't show up for work. Or they show up two hours late. Or, when he leaves the site, he returns to find they have indeed cut the rafters, but all too short. Regularly he has to fire somebody, then hunt for a replacement. The project, which could have been a four-month job, takes seven.

With a stable crew, you spend your time building and not hiring—a time-consuming task when you do it right. Your crews know your policies, how to get along with you, and how to treat your clients. When your clients recommend you, they are also recommending your crew. This gives you a big advantage over builders who can't tell clients who will actually be doing the work on the property. In addition, the people on the crews know how to work together. For example, the lead knows the apprentice's limits, which tasks to assign and how much explanation to give. The apprentice in turn knows the lead's shorthand—that a slash alongside an arrow indicates stud placement. Most important, stable crews work safely. In the state of California, for example, over 40% of injuries occur to workers who have been with their employers less than a year.

To create stable crews, you must pay well, and on pp. 185-190 I will discuss the varieties of pay you can offer. Workers, however, are also concerned with having their psychological needs fulfilled by their work. While a lot of "management psychology" strikes me as so much babble, there are a few straightforward human needs builders should consider as they work with their crews. Business consultant J. Mancuso summarizes them with down-to-earth succinctness. Employees, he says, have three fundamental needs other than a decent paycheck: predictability, control and recognition. (You will probably recognize the identical needs in yourself as an employer.)

Given the boom-and-bust cycle in our industry, predictability is especially important. As a small-volume builder, you can't do much to humanize the economy. But you can cushion your employees against it. Quite simply, you can let them know if you are heading into a slow period, and do what you can to help them bridge it. You may fear that if you tell your crews you are going to be slow, they will jump ship, leaving you shorthanded in the middle of a project. In my company, I have found a solution—helping my crew line up side jobs (small repair projects that the company cannot do cost effectively), for which they may use company tools. Because I record and file every single job prospect, no matter how small, that comes our way, we typically have no shortage of side work when we need it. Alternatively, and at times preferably, I can line up fill-in work for my people with another builder or through our temporary agency. In response, my employees have, virtually without exception, been candid with me about how far into the future I can count on them and have fulfilled their commitments. In other words, they have returned predictability with predictability.

Along with reliable employment, workers need emotional predictability in their employer. When I was a carpenter, the contractors I worked for frequently showed up at the job site in a dither. (One particularly crusty fellow, for whom I framed custom homes on steep lots, often began bellowing at us from the curb and receded into quieter contempt only as he reached our level: "You so-and-so's, you'll never be carpenters. How many times have I told you....") Now that I'm sitting in the same place as my old bosses, I can guess at the source of their rancor. They were driving out from their offices, where they had just taken a look at their job costs and realized they were perilously over budget. When they arrived at the project and saw work going more slowly than expected, they blew up in frustration and fear. But their behavior was counterproductive. It made workers feel so threatened that they became more concerned about looking busy than about actually getting work done.

Since becoming a builder myself, I have learned a few tricks to even out my behavior with the crew. If I feel edgy as I head for a job site, I detour to the nearest cafe, where I sit back with a cup of coffee until I settle into a calmer perspective. (What's money anyhow? The trees and sky and earth will still be here even if we don't make our profit margin on the job, and after it's done, we'll move on to another opportunity. And so forth....) When I reach the job, I try to greet everyone with a friendly hello. If I'm still tense, I admit it, and emphasize that the crew should not take my mood personally. When you are the boss—the source of the paycheck—you are an important figure in your employees' lives. Your irritability can hit them powerfully.

Control, the second of Mancuso's requirements, becomes especially important as you make the transition from independent tradesperson to running your own company. You must learn to delegate authority. If you don't, your workers will not be able to make the decisions needed for efficiency. But giving away control can be hard. One highly skilled carpenter who became a contractor recalls that when he first hired help, he tried to create a "bunch of clones, extra sets of his own hands." But then, he says, he began to see that he was "taking the enjoyment out of the work for the employees and killing their incentive." Finally, when his top lead walked off a job, he realized he had no choice but "to let go."

Along with control over their own work, employees need the chance to contribute to overall company policy. Ideally, you'll have constant give and take with your crew. If not, schedule regular feedback sessions. As stressed on pp. 31-35, never make major changes in your practices without crew input. I learned my lesson a few years back when I abruptly announced to my crew that henceforth I would provide only the larger power tools. They would be expected to bring to the job a long list of smaller power tools as well as hand tools. But with a combination of dark looks and candid objection, they quickly made it clear to me that such a policy, which could have a big impact on their finances, had to be thoroughly discussed before implementation.

Among the best ways small-volume builders can delegate authority is to let their crews take projects from tear-out through the finish punch list. That's not always consistent with competitiveness, especially on large jobs where subbing to contractors who specialize in foundations, frame or finish may be much more cost effective. But when a crew controls a job from start to finish, its members feel less like cogs in your production machine. The project is theirs in a way it cannot be when they have been only one in a string of carpentry and subcontractor crews run through the project at your behest. They will bring their friends by to see it and send photos to the family, announcing with pride, "I built this place." They are on their way to filling the last of the needs Mancuso spotlights.

Employees require that third item—recognition—in frequent doses, as I'm occasionally reminded when I forget to give it to my own crews. A top carpenter once responded with astonishment when I complimented him on his skills. In turn, I was flabbergasted, for I thought my admiration was obvious. But it had been a month since I

Company Organization

The general contractor may be the hub of the company, but at the job site the crew leader is the hub and the contractor a spoke. The contractor's job is to support the lead.

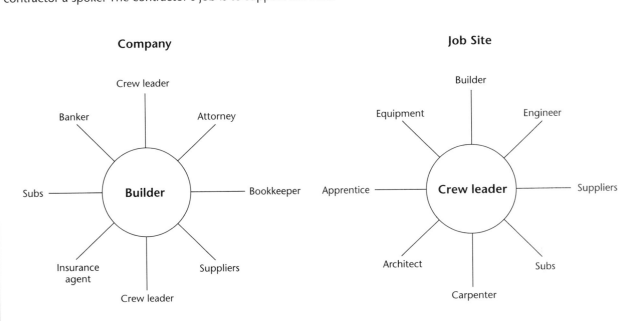

had expressed it, and he felt the absence of explicit appreciation as implicit criticism. Human confidence, as you can observe in the struggle of even great athletes to keep theirs intact, is fragile and needs constant bolstering.

A most important means of extending recognition to your employees is by promoting from within the company. Bringing in an outsider to occupy a rung of the ladder above existing crew members can be demoralizing and stir resentment. If you have a lead position open, by giving it to a top carpenter, even if she or he will need extra support for a time, you get the bonus of strengthening your company's morale.

You also give or deny recognition in the way you handle mistakes. It is an old management rule that you don't attack a person when you talk about a mistake. You criticize the error, not the individual. One builder begins discussions of errors with a phrase like, "How did it happen that a person with your skills...." He wants to find out how the mistake happened so as to prevent a recurrence, but also to leave his employee's confidence intact.

I'm dubious about the value of criticizing mistakes, even in carefully conditioned language. Capable tradespeople usually have figured out why they've made an error well before you are even aware of it, and they are much tougher on themselves than you can productively be. For people who make frequent errors, you can't do much with criticism. They need to be in another line of work, and you have to help them find it. As a result, I find it most useful to downplay individual errors, even serious ones, and instead to concentrate on overall performance. To improve it, I offer insight into bad habits (such as apprentices dropping their bags when they don't need their tools for a moment or leads relying on memory instead of lists), which, unlike errors, are usually invisible to the perpetrators.

CREW LEADERS

Good crew leaders are masters of one of the toughest jobs in construction. Like builders, they must provide predictability, control and recognition to other workers—and often to their boss, as well. At the same time they must tightly control the work, maintaining square, plumb and level. In other words, they must simultaneously maintain two opposed attitudes. They must be strict with their material. But they must be flexible and accommodating with their crew. And they must know when to reverse those attitudes—to yield to the material a bit and to clamp down on the people.

Fred Blodgett, an ace crew leader who started working with Gerstel 13 years ago and is as sharp with pen, paper and phone as he is skilled at carpentry and crew leadership, assembled this compact center for administering his projects.

The specific duties of your leads will depend on their skill levels, the size of your company and the complexity of your projects. At the most basic level, leads must know their trade thoroughly and be able to read plans and specifications and to perform accurate layout. They must be able to lead a crew in executing the work, and coordinate with subs as they arrive at the project. Even at the basic level, leads must be able to communicate clearly, completely and concisely. I have known some who grunted their instructions; others offered long, baroquely elaborate explanations. With both types, workers struggle to figure out what they are being told to do.

As they move up a level, leads become responsible for arranging the delivery of material to the site. They take a big jump, for they will need to think ahead not merely a few hours, but days and weeks

A Crew Leader's Duties

The responsibilities of leads can be grouped into three levels:

Basic
- Read plans and perform layout
- Perform any carpentry task as needed
- Perform minor subtrade tasks on occasion
- Lead and supervise laborers, apprentices and journey-level workers at carpentry and associated tasks
- Coordinate carpentry work with subtrades
- Receive and check material orders
- File job records, including invoices
- Ensure that the job site is clean, well organized and safe
- Communicate with customers courteously and direct their questions to the proper individual, whether designer, engineer or contractor

Intermediate (additional items)
- Develop alternate and more efficient means of production
- Order materials
- Coordinate subcontractors at the site
- Check sub work
- See to accurate completion of time cards
- Call for inspections

Advanced (additional items)
- Schedule subcontractors' work
- Work out design problems with designer or engineer
- Deliver invoices and collect progress payments according to contract
- Produce estimates for change orders
- Write change orders and get clients' signatures
- Do final walk through with client and/or designer and create final punch list

at time. To avoid costly bottlenecks and runs back and forth to supply yards, intermediate-level leads must anticipate when they will need their material and make sure it is brought to their job safely in advance of that time. In my company, I found that delegating material ordering to my leads added much to overall efficiency. First, it saved a lot of my time. Second, doing take-offs allows leads to develop detailed knowledge of the plans, which they need to work effectively. Third, leads are able to work more quickly from their own take-offs than when they must match mine to the phases of work shown on the plans.

At the most advanced level, crew leaders take over scheduling subcontractors at the site. With material ordering and subcontractors under their control, leads assume complete command of a project's flow. As a last step, if your company grows to the point that you can not regularly make it to the job sites, you may wish to delegate change-order writing to your leads. Do so only with extreme care. Inadequate change-order writing is a prime cause of financial disaster for construction companies.

Change-order writing incorporates estimating, and, like other green estimators, leads tend to leave out many costs. That's if they even write the change orders. Because of the resistance they will feel from customers, and because they do not want to be distracted from production, they will often neglect to write them altogether. If you have succeeded in merely training your leads to notify you of all extra work requiring change orders so that you can write them up, you have accomplished much. Should you wish to delegate the actual writing of the change orders, you must provide leads with an estimating checklist, teach them how to use it, supervise their change-order writing closely for several projects, and monitor it thereafter. Stamp into their consciousness that the financial success of their projects, and therefore their employment, depends on their keeping up with change orders. Even if it means taking heat from clients, they must avoid giving away changes, and thereby the company margin for overhead and profit, to clients.

Crew leaders must be leaders. Along with their tangible duties of building and organizing, they have responsibility for crew morale. I have known leads who were excellent at the technical tasks—skillful with tools, adept at plan reading and layout and meticulous at scheduling. They communicated with precision. But they were still lousy leads. Some were moody, and their glumness made their crews hate coming to work. Others were greedy. They hogged all the interesting hands-on work. Or they rarely chipped in with the dirty work and treated their apprentices like servants. Some were so harsh and constant in their criticisms that their crews worked at a snail's pace for fear of making a mistake. Others so needed to be liked that they could not make the necessary demands of their workers or firmly direct young, headstrong apprentices. As a result, productivity dragged on the projects run by these highly skilled people.

Good crew leaders buoy their crew's spirits and foster enthusiasm for the project. They are patient, even-tempered and have a sense of humor. (One way to tell whether morale is good at a work site is to listen for laughter.) They know how to support their workers with praise and correct them without dispiriting harshness. At the same time, they can push workers with a gruff good nature that unmistakably says, "Move it or else." I have tremendous respect for top-notch leads. They are rare. Managing them is one of the great challenges of being a builder, for you are managing leaders.

When you put individuals in lead positions, you must define their responsibilities clearly. And once they are established, leave them in their positions. Don't bounce a lead back down to a carpenter's role, then up again, then down. One of my own mistakes is illustrative of the damage you can do. During a major project, the lead took a week's vacation. While he was gone, I gave his duties to a carpenter whom I was readying to take on his own crew. So far so good. But then I decided that for the sake of continuity, when the lead returned, I would assign him to build a complex staircase for the project. He had expressed interest in the task, and I figured that while he built the staircase the carpenter could continue to lead the job and hone his project-organization skills. When the lead heard my idea, he told me candidly that he felt humiliated, as if he had been stripped of rank in front of the crew and subcontractors he had been leading for months.

To do their jobs effectively, and to improve their technical and leadership skills, crew leaders need the authority to run their projects. You can't take back control when it suits you. One capable lead told me, "If I'm in charge, I don't want him (the builder) there. He comes to the site and goes off half-cocked. 'No, no, no' he's yelling, but we have our systems all worked out, and we don't want to be stopped in midprocess."

You must recognize that while you still write the checks, the lead's crew is his or her own. If you send out instructions to the workers directly, you undermine the lead's authority, and give the message that it's okay to bypass the lead. Crew members may start coming to you with their problems. In so doing, they may be subtly registering their disapproval of the lead, and by letting them, you are sanctioning it. You must back your lead even when someone on the crew comes to you with a legitimate complaint. One skilled manager recommends listening respectfully to such complaints, neither denying or confirming them, then suggesting that the problems be raised directly with the crew leader.

If the crew members complain persistently, intensely and for good reason, or quit in resentment, you have a tougher row to hoe. You'll have to talk to the lead and point out that success at the job includes filling the crew's psychological needs and winning their loyalty and support. You may have to fire a lead if, with time, he or she does not learn to handle the authority productively. You cannot keep feeding even the best carpenter and job organizer fresh bodies. It's wrong, and it's too expensive.

 # PAY

A former president of L.L. Bean, the extraordinarily successful mail-order clothier, held that "Paying 20% above average in wages will get you a 30% to 40% above-average employee." If you pay well rather than squeeze your workers, they in turn will not feel compelled to hold back on you. I would, however, qualify the L.L. Bean axiom: When you hire new people, especially apprentices, find out what they are worth before loosening the purse strings. Set a probationary pay level, if only for a few days, and a target wage. After a trial period, you can increase pay if the new employee seems like a good candidate for a longer run with your crew. At regular intervals, talk with your permanent employees to agree upon a schedule of raises. Their pay should rise both to keep up with inflation and in compensation for increases in skill and efficiency.

Along with a good hourly wage, good pay can include many additional benefits. In the world of small-volume construction, where "under-the-table" arrangements are so common, the first additional benefit can and should be a legitimate paycheck. In my experience, tradespeople generally do not like the risk of concealing sizable amounts of under-the-table pay from the IRS. They also resent having to make the estimated tax payments and higher Social-Security contributions that go with reporting cash wages. In addition, they value the disability, unemployment and worker's-compensation insurance that accompany a legitimate paycheck.

You will want to show your employees the value of the benefits that go with the legitimate paychecks you provide. For example, the annual premium withheld from my crews' paychecks for disability insurance is about $200—far less than the cost of a private policy. Social-Security contributions, unemployment insurance and worker's-compensation coverage, which are paid out of the company account, run about $10,000 annually for our top carpenters.

___

As your company gains financial strength, you can provide additional benefits. Do bear in mind that each new benefit entails tax consequences, which you should discuss with your accountant. Make certain, also, that when you extend benefits you will be able to sustain them. If you enjoy a boom year early in your career and lavish benefits on your employees, you may find you will have to pull them when business dips. As a result, employee morale can also plummet. (As a wise manager once told me, "You can giveth but not taketh away.")

To ensure continuity and good morale, begin with less costly benefits and gradually work up to the more substantial ones. Three of the least burdensome benefits commonly granted by builders are really reimbursements for employees' out-of-pocket expenses:

- Commuting costs. Many builders pay their employees a per-mile allowance and travel time for projects beyond their usual range, such as 20 miles or half an hour.
- Tool allowances. Some companies pay their employees a flat fee— hourly, weekly or monthly—for use of their tools on company jobs. Alternatively, you can cover employees' costs for tool maintenance, repair and replacement. In addition, you can repay the use of employees' tools on your projects with use of company equipment for their side jobs.

Benefits

Benefits should be introduced slowly, so as not to overburden the company or unrealistically raise employees' expectations. Here is one logical order for establishing benefits beyond a legitimate paycheck:

- Commuting costs
- Travel time
- Repair and replacement of personal tools used on company projects
- Use of company tools for personal and side jobs
- Small power tools (pat-on-the-back bonus for apprentices)
- Tool allowance
- Vehicle allowance
- Health plan
- Profit sharing
- Paid holiday(s)
- Dental plan
- Pension plan

- Vehicle allowances. If employees use their vehicles for company business, such as material pickups, you can compensate them with a per-mile payment.

After reimbursements you can move up to paid holidays. These are a treat for employees and place little load on your company. Much of the usual labor burden is absent; since the employee is not actually working, you do not have to pay for crew supplies, worker's compensation or liability insurance. Thus, you can give an employee whose wage is $15 an hour a paid Thanksgiving for about $135.

As your company matures, your employees may come to expect, and you will feel ready to provide, the major traditional benefits—medical coverage and a retirement plan. Employees often prefer these benefits to wage increases because they do not have to pay taxes on them. For journey-level people, whose wages may put them in a high tax bracket, a tax-free benefit can have nearly twice the value of a wage increase. For example, when top carpenters' wages rise $100 a month, they will see only about $60 after taxes. But if they receive $100 worth of medical insurance, they get the full value with no taxes withheld.

The major benefits are expensive. As of 1990, depending on your location, medical insurance for each employee will likely run at least $100 monthly, or $6,000 annually for a crew of five. If you are covering employees' families as well, it can cost several times as much, and the cost will probably rise rapidly. It has for years and no slowdown is in sight as of this writing. If necessary, you can moderate your costs by asking employees to pay for part of their coverage. Note that if you do elect to provide health coverage, you have alternatives to offering your employees participation in a group plan. In my company, for example, each employee obtains his or her own health coverage. I then write checks to the various providers—not to the employees—each month or quarter as required. My costs are nearly identical to those for a plan for a small group. And if employees move on, they can pick up the payments for their coverage if they wish.

A fully equipped modern carpenter has made a large investment in tools and should be reimbursed by one means or another for their use on company jobs.

With pension plans, the burden can be not only financial, but also bureaucratic. For most plans, the paperwork is more than small-volume builders will want to deal with. Fortunately, there is a plan available that has deliberately been simplified for use by small businesses. Called the Simplified Employment Pension (SEP), this plan allows tax-free contributions to be made into employees' Individual Retirement Accounts (IRAs). Here is how the SEP works:

- Employees open their own IRAs, typically at a bank or brokerage house. All they need do is fill out a short form and send it in with a few dollars. Now they have an IRA. If they wish, they can make direct contributions to it. Interest or other earnings on the account are tax free until drawn out upon retirement.

- You make SEP contributions directly to your employees' IRAs. The employees then assume full control of investing the funds (stock market, bonds, money market, etc.), just as with any other IRA contribution.

- If you wish, you can make a contribution to your own account that is equal (as a percentage of earnings) to that you have made for each employee.

There are restrictions on SEPs, the most important of which is that if you make a contribution for yourself, you must also contribute for all employees who have worked for you more than three years. With my own company, however, I have found (in good years at least), that I can make an SEP contributions of 12% to 15% to my own and my senior employees' accounts and still break even. In other words, the contributions to my employees' accounts about equal the additional taxes I would have had to pay if I'd made no SEP contributions at all. The employees get a big tax-free bonus. I transfer a sizable chunk of tax-free cash to my own retirement account (where it begins earning more tax-free dollars). And I don't end up paying any more money out-of-pocket than if none of us had gotten an SEP contribution. Great stuff!

I get extra mileage from the SEP-IRA program by using it as a profit-sharing mechanism. Contributions to employee plans are linked to company earnings. As the policy statement explains to employees, contributions begin only after the company has covered all overhead, including compensation for my management work. To illustrate: For 1993 I could determine that SEP contributions will begin after the company earns enough to cover all job and overhead costs, including $55,000 for my pay. For each additional $3,000 in earnings, eligible employees will receive an amount equal to 1% of their gross wages as a SEP contribution. Contributions will top out at 15%

of wages, the legal limit. Therefore, if the company takes in $45,000 beyond all costs, eligible employees get 15% of their wages as a SEP contribution—a carpenter who has received $30,000 in wages will thus receive a $4,500 SEP contribution.

It's up to you to determine the point at which the SEP kicks in and the rate at which the benefit to your employees increases. You can vary your formula to fit changing conditions in your company. To continue the preceding example, in 1995 you might determine that you deserve a raise, and that employee SEP benefits should not kick in until you have received $60,000 in compensation. Or you might see that because more employees are eligible, you can only afford to pay out SEP contributions at a rate of 1% of wages for every $4,000 in profit instead of $3,000.

The SEP as a profit-sharing device—rather than a guaranteed pension contribution tied directly to wages regardless of profit—has an additional attraction for a small-volume company. It does not excessively burden you in a slow year. If the company experiences marginal financial performance during an economic recession, SEP contributions will be minimal—or even zero, if the company doesn't earn enough to cover your salary. You will not be loaded with responsibility for an expensive benefit your company can ill afford. On the other hand, when business is strong and the company prospers, employees justly share in the profit they have helped to create.

Profit sharing increases employee loyalty, and, therefore, crew stability. It gives employees a more direct financial stake in the company, and encourages them to develop a concern for its long-term performance. In this respect, profit sharing may well be superior to the use of bonuses handed out a project at a time. When employees are promised a bonus for pushing a particular job to speedy and profitable completion, they are encouraged to cut corners for the sake of the quick return. But longer-term profit sharing, by tying income to the overall performance of the company, can encourage a longer-term perspective. It promotes the understanding that a company's prosperity is based on doing each project well and leaving behind a satisfied customer who refers others looking for good work. Profit sharing is an enlightened form of self-interest all companies could usefully incorporate. The SEP is a mechanism by which small-volume construction companies can provide a pension plan and at the same time practice profit sharing.

THE FOUR-DAY WEEK

Pay can take less tangible forms than good wages, medical plans and SEPs. Builder John Larson describes some of the other important considerations: "My crew knows they can make more out at some tract just slapping something together with a bunch of kids snorting speed. But the work we do is not dumb work.... If you offer a good job in a humane environment, a chance to really practice the craft, that's definitely worth something."

A "humane" environment in construction includes a site that is clean, safe, well-organized and worked by courteous people. It can also productively include a schedule better than the customary five-day week. In my company, we work a four-day week, then take off three. Work begins daily at 7:30 a.m. and ends at 5:45 p.m. With a 45-minute lunch break, the crews put in 9½ hours paid time each day, or 38 hours a week.

The "4/3," as we call our schedule, has benefited my company so greatly that I have come to advocate its wider adoption in the industry. Before describing its advantages, however, I should mention a few potential drawbacks. The 4/3 does not work for some employees. They are unable to manage so much free time. One builder I know wanted to change to a 4/3 but was prevented from doing so by a top carpenter who said he'd quit. The man was terrified at the prospect of having to fill three whole days on his own initiative. Other employees find the 4/3 interferes with their child-rearing responsibilities. The necessary early start to the workday can make it difficult to get small children off to school.

The 4/3 can also add cost to a project for artificial lighting. During the short winter days, the sun may rise after the start of the workday and set before it is over. On spread-out projects, lighting could possibly require a major effort. In my company, however, on large projects we adjust our starting time to correspond to first daylight; by setting up a few lamps, we are able to remain efficient with little cost.

On physically demanding projects, such as large framing jobs, the longer workday might cause exhaustion, and, consequently, inefficiency and danger. That has been a concern of other builders to whom I have touted the 4/3. Again, my own experience suggests the concern is exaggerated. Although most of my company's jobs involve much heavy labor—we regularly build large concrete forms and frames—we have not found exhaustion to be a problem. Our projects are built safely on tight schedules, and we're quite competitive in price with builders who produce work of comparable quality. With morning and afternoon breaks (as are also legally required for an eight-hour day), the crews easily handle the longer workday. In fact, when one of our carpenters worked for another builder during a lull in our schedule, he reported that after years on the 4/3 he found a five-day week exhausting! Early in my career, I worked a seven-day/12-hour schedule (followed by seven days off), and I loved it. It's a matter of what you're used to, I suspect. The exhaustion barrier may be largely psychological. Five-day workers seem to reach it Friday afternoons.

If the potential drawbacks to the 4/3 are minor or exaggerated, its advantages are numerous, beginning with a couple of increases in efficiency: First, daily setup, rollup and cleanup time is reduced by 20%, since each task is performed four and not five times weekly. Second, less time is lost to holidays. In 1989, for example, both Christmas and New Year's Day landed on Monday, an off day for our crews, since we work Tuesday through Friday. They got in their full work week while still enjoying long holiday weekends.

Clients, especially those with remodeling projects, like the 4/3. Along with their neighbors, they have to endure the noise and dirt of construction one day less a week. Subcontractors, too, can make use of the 4/3 schedule. If they need a carpenter on hand, they come to the site on one of the four workdays. But if they prefer, they can come in during the off day and have the site to themselves.

As a contractor, I have found that having my projects active for four days a week instead of five greatly consolidates my work load. Indeed, I can often compact the work of running my company into three days. Wednesday and Friday I visit job sites and prospective clients. Thursday is for job costing, estimating and other office work. The remaining days of the week I may choose to do a few incidental chores in the morning. But afterward I am free to do consulting, write or shoot hoops.

To compensate for the lack of light at the beginning and end of short winter days, Gerstel's crews set up a few banks of lights. They are so helpful on interior jobs that carpenters generally leave the lights on all day.

But the greatest advantage of the 4/3 is the attraction it holds for employees. When I advertise for workers at any level, I often get responses from people with good jobs in other companies. Several times apprentices and carpenters who had their pick of jobs have elected to come with us. The 4/3 played a large part in bringing these people our way.

Once they have joined us, the carpenters like the freedom afforded by the 4/3 as much as I do. One of them told me that with the 4/3, he feels as if he works only half a week. (In fact, that is the case. From the time work starts Tuesday morning until it ends on Friday is only 82 hours, less than half the 168 hours in a week.) With the 4/3, crews can adjust their schedules so that weekends, especially holiday weekends, can turn into mini-vacations with little loss of time and pay. For example, at Thanksgiving, which lands on a Thursday, the crew shifts their schedule and works the preceding Monday, Tuesday and Wednesday. By returning to work the following Tuesday, their normal starting day, they enjoy five days off with the loss of only a single workday.

Out of the advantages to the employees comes a great advantage to the company—improved crew morale. We don't get much "Thank God it's Friday" or "Oh hell, it's already Monday and I have to go back to work." The four days of work pass quickly because there is always a long weekend to look forward to. And after it's over, the crews feel refreshed and ready for work again.

Because the carpenters like it so much, our use of the 4/3 minimizes employee turnover. Turnover is exceedingly expensive. One study shows that each time a position in a company is filled, hiring and training the new person generates costs equal to 20% to 100% of his or her annual salary. Anything you can do to diminish turnover greatly strengthens your company for the long run. Recently, a contractor I know put his crew on the 4/3. After a month I asked him how it was working out. He answered with no hesitation: "We'll never go back."

 # SUBCONTRACTORS AND SUPPLIERS

A single subcontractor performing poorly can throw off an entire project. Suppose, for example, that the plumbers take on more work than they can handle and as a result arrive at your project two weeks late. Because their pipes are not in the walls, you can't place your insulation batts. As a result, the drywallers can't hang their rock. By the time the plumbers have finally gotten in and out, the rockers have moved on to other commitments, and you can't get them to your project for three more weeks. Meanwhile, you have to accept delivery of cabinets and store them on site, where they will be in the drywallers' way. When the drywallers see the cabinets, they get mad, and either refuse to work or raise their charges. And so on, all because one subcontractor got behind.

Because of trouble with unreliable subs, many small-volume builders try to have their crews perform all trades. In some cases, even when good subs are available, handling their trades "in-house" may be best. If a project requires only a small amount of work in a subtrade—running one or two electrical circuits or changing a plumbing fixture—subbing it out is likely not worth the management effort. Sometimes you will have your crew do even a substantial volume of work in a subtrade in order to keep them employed. For example, on kitchen remodels, though it costs us more, my crews frequently install and tape the drywall, since otherwise there would be no work for them at the site while it went up.

Such exceptions aside, subcontracting out the specialty work—or even some carpentry, such as a complex stairway—is normally more cost effective. Of course, you may sometimes want to handle a subtrade or an unusual carpentry item out of sheer fascination with the work. But be aware that your reward will be education and the pleasure of ex-

A well-equipped subcontractor has a large investment in specialized tools and supplies, which a general contractor who uses them only occasionally likely will not find cost-effective.

tending your skills. Financially, you will probably turn in a marginal performance, or even take a beating. The economic advantages of subcontracting are many, beginning at estimating and extending beyond a project's completion:

- Becoming expert at estimating for general conditions, rough carpentry and finish carpentry is challenge enough. When you try to develop the knowledge and maintain the unit-cost files to also estimate the other trades you will likely leave large openings for underestimates and cost overruns.
- Every bid entails risk. When you subcontract portions of a project, you spread the risk.
- Subcontractors are likely to be cheaper. Their materials typically cost less because they purchase larger volumes from their suppliers than you would. Their labor rates may seem high compared to what your crew costs you. But when you factor in their greater speed as well as the costs you avoid (extra management responsibility and additional overhead for specialized equipment), the spread disappears. Or it swings in the subs' favor.
- By subcontracting, you can increase both client satisfaction and earnings. Instead of your crew doing rough carpentry, then sheet metal, then plumbing, then electrical, and so on, the subcontractors perform their trades even while your crew pushes carpentry forward. As a result, the project is completed more quickly and you are able to move on to the next. Over the course of the year,

subcontracting may allow a crew to do several additional projects, with that much more opportunity for earnings.

- Subcontracting portions of your projects will enable you to maintain smaller, more effective crews.
  - Subcontractors expert at their trade can achieve higher quality than your employees, who handle it only a few times a year.
  - After a project is complete, subcontractors share with you the burden of responsibility for any problems that might develop.

Successful builders create mutually supportive relationships with skilled subcontractors. Moses Brown, owner of Kensington Plastering, is one of Gerstel's valued subcontractors.

To have a bid accepted "is not to win the race, but only to qualify for the real contest that lies ahead—the construction," Paul Cook reminds us in his fine book on estimating and bidding, *Estimating for the General Contractor* (see Resources, p. 223). You will not win that contest, he adds, if your subs are of the "weak, low-bidding variety." In the following pages, we will discuss finding strong subcontractors, what you must do to keep them and what you can expect in return.

When you contract with a sub for the first time, you should exercise the same care you do in hiring a new employee. A good source for recommendations are local building inspectors, who know which subs in your area consistently produce sound work. Sometimes you can spot promising subs at job sites around town. But the best and most direct way to find subs is to call other builders.

As the checklist at right suggests, you want subs who can deliver the same dual capabilities you cultivate in your company: skill with the tools supported by strong management. You can learn much about a prospect with an initial phone interview—the more conversational the better (see p. 172). Afterward, check with the sub's general-contractor references, comparing the sub's claims against actual performance. Check the sub's credit with suppliers. If credit is poor, and the sub is unable to procure material in the midst of your project, it can be brought to a halt. You want subs who have had good credit for at least a couple of years. If their accounts have been open for less time, hesitate to use them for any but small projects.

When you qualify a sub, you need also to qualify the crew. You want to be sure that the same people who did the work for the sub's references will be at your project. Try to take a look at a sub's work, preferably at several projects and while the crews are working, so you can ascertain whether you, your crew and clients will be comfortable with them.

Qualifying a subcontractor is not a one-shot procedure. Unfortunately, a good sub can quite suddenly turn into a dud. You must keep an eye on your subs, talk with them about their operations and talk about them with their other general contractors. Watch out for turnover. A sub who is unable to hold employees will not be consistent. Be especially alert for overextension. During a strong economy, the best subcontractors can become inundated with work. Flattered, tempted by greater profit, eager for the bragging rights of being "big," they hire too many new workers too quickly and lose control of their operations. Both the quality of their work and their reliability plummet. Several times subs who had given us good performances for years suddenly overexpanded and did a lousy job on our next project.

Once you have found good subs, you want to keep them. Over the long haul, you will be much better off than if you "churn" your subs in pursuit of the lowest bid on every project. Good subs can afford to turn away your business when you need them most, especially during a strong economy. To keep them, you must do several things: treat them with courtesy and respect, organize your projects so that they can get their work done efficiently, pay them promptly and accept their fair-market bids.

Your subs require from you the same respect you want from your clients. They do not want to be treated like lackeys or taken for granted. I often refer to my subs as "specialty contractors" or "co-contractors," to remind myself that, while legally speaking their subcontract is subordinate to my prime contract with the client, they are not underlings. They are equal players on our team.

Once construction of a project begins, be attentive to your subcontractors' needs. For starters, give them ample notice of when you will need them at the job site. Steve Nicholls, owner of a highly respected cabinet shop, explains, "You are not your subs' only client. Don't act like you are. Take care to give them the time they need to effectively work you into their schedule." Before you call in your subs, make certain that your project is ready for their crews to work. Few things anger subs more than being asked to mobilize a truck and crew, then arriving at the site to find the project has not progressed far enough for them to work efficiently. Electrical contractor Hagen Finley describes the results: "I've wasted a good part of my day, and I'm out several hours of wages getting my crew out to the job. Then I

Qualifying a Subcontractor

A checklist helps you to evaluate subcontractors you are considering using for the first time:

• License

• Worker's-compensation insurance

• Liability insurance

• Number of years in business

• Stable crew

• Credit standing with suppliers

• Quality of rough work

• Quality of finish work

• Niche: Custom, standard or production

• Size of job you handle

• Adequate crew

• Is overextension a problem?

• Do they begin on schedule?

• Do they wrap-up on schedule?

• Cleanup procedures

• Success with inspections

• Cooperation with other trades

• Relationships with leads and other crew

• Relationships with clients

• Responsiveness to problems

• Availability (answering machine, voice mail, beeper)

• Change-order practices

• Promptness on callbacks

• Relationship to reference (How long have you worked together? Do you like him or her?)

have to move the crew to a second job. But because we get started on it late, we can't get it done in one day. So we have to return a second day, incurring redundant travel and setup time, all due to the contractor on the first job telling us he was ready when he wasn't."

To make sure your site is ready, you can ask a sub to inspect it before bringing on a crew. Some builders get double duty from such inspections by asking for a formal acceptance of the work of each trade by the succeeding one. The drywallers are asked to accept the framing that must take their rock. The painters accept the drywall finish. The roofers accept the flashings and so on. Then, in the event of later problems, a sub cannot attempt to deflect blame to a predecessor. Once subs are at work at your site, give them the support they need to work efficiently. Your crew should promptly do any needed carpentry, but otherwise attempt to stay out of the subs' way.

When their work is completed and they have billed you, pay your subs promptly. Speaking of a successful builder, Steve Nicholls said, "I just built $15,000 worth of cabinets for him. He sent me a check in a week and along with it a nice thank-you note. What will I do the next time he calls? Why, he will get my immediate and careful attention, of course."

Finally, be willing to accept fair-market bids from your subcontractors. You don't want cheap subs. After callbacks, lawsuits and the bitter clients they bring down on you, they will not be cheap. Of course, you expect your subs to be competitive, and you should probably have several who want to bid for your projects. Occasionally you will take a bid from an entirely new person, who may then win a place in your regular rotation. Now and then you can ask your subs to sharpen their pencils to help make a project possible. But do not grind on them job after job. And do not squeeze them for extras after a project is underway. When one of our electricians cut another builder from his client list, he explained, "The guy was always whining. He was always asking for extras and concessions. He killed my profit on every job."

Avoid the sleazy practice of "bid peddling." Do not, for example, take a bid from Scoll Electric and then ask Alpen Electric to beat it. Bid peddling degrades the construction industry by pushing subs to work without adequate margin for risk and profit and, as a result, to cut corners. Bear in mind that subs who are willing to peddle their bids and shaft their competitors will probably also shaft you. Reputable subs will not put up with bid peddling. "If I ever find out a general contractor peddled one of my bids," says my sheet-metal sub, "I'll never give him another price."

If you stick with your subcontractors, treat them thoughtfully, and pay them promptly and fairly, you can reasonably expect much in return:

- Full service. When your subs are overloaded, it should not be your project that gets pushed to the back burner. They should be willing, also, to take care of the little nuisance jobs, because they know they will be getting your better projects, too.
- Clarification of incidental responsibilities and conflicts during bidding. Good subs cooperate with you in determining which overlap responsibilities (such as demolition and cleanup) will be handled by their crew and which by yours, so that no work goes unaccounted for in your estimate. They should look for potential conflicts between their work and that of other trades. For example, if a structure is heavily engineered, the plumbers should check whether they will have room to run their waste lines without compromising the integrity of posts and beams.
- Advance ordering of any custom materials requiring long lead times for delivery to the project. If material is late and bottlenecking the job, subs should go to whatever necessary expense, including paying air freight, to get it to the site and installed.
- Timely execution of work. Subs should get to your project on time and with an adequate work force to accomplish their work on schedule.
- Prompt response to callbacks. If problems develop with their work after completion of the project, good subs return to make adjustments or repairs immediately.
- Prompt billing. They get an invoice to you quickly, but do not badger you for instant payment, instead allowing you time to put their bill through your usual bookkeeping procedures.

If you are giving subs what they need and are not getting top performance in return, you will find others eager to take their place. A plumbing contractor who solicited our business explained, "There are a lot of lousy subs. But there are also a lot of lousy generals out there that can ruin a sub. I'm looking for a few good ones."

While subcontractors get a lot of ink, suppliers are hardly mentioned in books on construction management. The imbalance is odd, because good suppliers are as critical to a builder's success as reliable subs. A construction project can require a confluence of material from all over the country, or even the world. The suppliers bring that material to you. A poor performance—late delivery, wrong delivery, no delivery—causes serious problems. To illustrate, before we contracted for the job, a customer for a recent kitchen remodel had ordered her hood from a supplier I had never dealt with. The supplier first sent the hood with the wrong duct cover. Six weeks later the correct cover arrived, but it had been crushed as a result of improper packing. Back it went. An intact cover finally reached the job weeks after all the other work was complete. Three workers had to return to the job site to put in the hood and cover, costing us considerable travel and setup time.

Factors to consider when selecting suppliers are:

- Condition of the supplier's yard and/or shop. What suppliers really sell is organization and procedures that get you what you need when you need it. If the yard or shop is not well organized, it's not likely the procedures will be either.

  - Staff. Is it stable? A steady working relationship with a supplier's sales rep can greatly improve your efficiency. If turnover is high, you will constantly be working with new people, learning their styles as they learn yours and living with misunderstandings and mistakes.

  - Quality. Sometimes you can get more for less. But usually you get less for less.

  - Cost. On closer examination, that seemingly more expensive supplier may not be. For example, my favorite yard's retail charge for framing lumber is about 15% higher than a yard in the next town over. But with the greater discount my yard gives me, the difference drops to 5%. In addition, my crews must cull less material—the cheaper yard's stock is likely to include severely crowned, twisted and waned boards. The resultant savings of material and greater ease of building lower the cost difference to a negligible level. Factor in the far better service, more flexible return policy and lower delivery charge, and our yard turns out to offer a better price than the "cheaper" competitor.

- Billing. Itemized bills are a must. Without them, you cannot check the charges (which are wrong a significant percentage of the time) on individual items, nor can you accurately job cost. Computerized bills are helpful, because the chances of error are lessened.

Good work comes out of well-organized shops, such as this cabinet shop operated by Lon Williams.

Once you have established yourself with good suppliers, stick with them. They become part of your "team." You have a mutually supportive relationship with their sales representatives: You are patient with their occasional errors; they will do their best when you urgently need an item. They know your standards and come up with ideas that help you to achieve your goals. For example, for an addition my company built a few years back, I originally ordered 2x6 Douglas fir for exposed rafters as specified in the architect's drawings. My lumber-yard sales rep, knowing we would not be satisfied with the crude look of the fir, suggested we use kiln-dry hemlock instead. He had just gotten in a shipment of good-looking stock, and was able to provide it to us for only slightly more cost.

By sticking with a supplier, you can reap financial benefits. You may get a larger discount, since discounts are often based on overall volume, not the size of an individual order. In the long run, therefore, you may get better prices than by shopping suppliers for low bids on every project. Suppliers who view you as a loyal customer with a track record for paying the bills can be allies in time of cash crisis. Rather than cutting off your credit, they will stick with you.

As with subs, you can do much to create good relationships with suppliers. To begin with, pay them. General contractors are considered bad financial risks by suppliers. At their association meetings, a common question is "Which of your contractors went bankrupt this month, and how much did they leave you holding the bag for?" You can win your suppliers' confidence by paying promptly, year after year.

Give your suppliers respect. Often they are much larger companies than the general contractors they service. Yet they find themselves treated cavalierly. Regularly, builders barge in on their sales reps demanding instant attention, as if the reps had no one else to service.

Don't expect that your suppliers won't make mistakes, only that they will take care of the mistakes promptly. "Good contractors," says cabinetmaker Steve Nicholls, "stay calm about mistakes. They know they will be worked out." At the same time, let your suppliers know you appreciate their work. "We all need encouragement for what we do," says Nicholls. "I love working with people who like us. I love it. I will bend over backward to take care of those people who appreciate what we do."

PART 8 PROJECT MANAGEMENT

Safety
Job Setup
Running Projects
Inspectors
Wrap-up and Follow-Up

SAFETY

Achieving a good safety record is difficult in construction, which is among the most dangerous of industries. Some 600 people a year are killed in construction work in the U.S., and a huge number is severely injured. But if you do not manage your projects with safety as the highest priority—and I mean *the* highest—you can lose both the emotional and financial rewards of the work. Your negligence can ruin someone's life. The consequent increase in your worker's compensation rates can destroy the competitiveness of your company and your ability to earn a decent income.

A checklist can remind you and your crews of all the potential hazards and the necessary protective procedures. Of the items on the list shown on the facing page, a few deserve comment:

- Careful ladder use. Because the smaller ladders appear quite harmless, workers tend to use them without due respect. Human beings, however, are poorly designed for falling, and you don't have to fall far off a ladder to get badly hurt. One worker who slipped a few feet down a ladder smashed one heel into a rung, crushing it so badly he has since walked with a limp.

- Personal protection. Some protective gear, such as goggles, steel-toed boots and hard hats, prevents those dramatic, instantaneous injuries. A second kind of gear, like respirators, nose and mouth masks, knee pads and ear covers, prevents the injuries that come from a little damage done day after day over many years. But because these cumulative injuries seem remote, workers often ignore them. Ear protection is particularly neglected. As a result, a large proportion of tradespeople suffer partial or complete hearing loss. One smart carpenter I know wears ear plugs at all times, and keeps a headset on his tool pouch so that he can quickly slip it on whenever he operates or is exposed to a loud power tool.

- Power tools. Power tools are obviously threatening, and for that reason they command more respect than the quieter and more insidious hazards on a construction site. The scream of a saw or router insists that you take care, and most workers listen. Nevertheless, all new workers should be checked for correct use of all power tools. They should be made aware that tools can be dangerous when they are off as well as on. Consider this scenario: A carpenter descending a ladder is holding his framing gun by the trigger. He bumps its nose against the head of another carpenter at the bottom of the ladder and fires a 16d nail into his skull.

- First-aid kit. A minor cut can become infected for lack of antiseptic and a bandage. Every job site needs a first-aid kit, which should be restocked regularly. I procure first-aid materials by mail to save running back and forth to a pharmacy. The leads simply tell me when an item is running low, and I phone in an order during my weekly stint in the office.

But a truckload of safety gear will not create a safe work site without a strong commitment to safety on the part of both workers and management. According to the United States Department of Labor, worker carelessness and neglect of good safety practice, as opposed to

A job site should be well stocked with safety equipment and first-aid supplies.

Safety First

Ideally, a safety checklist should be posted at the job site. Important item include:

Regular job-site safety meetings
- Frequent, brief meetings are most effective.

First aid and emergency treatment
- Make sure the crew leader has the phone number for emergency services or ambulance services and the hospital nearest the job.
- Ensure the crew leader is trained in first-aid procedures, including CPR.
- Have a first-aid kit with supplies for washing, disinfecting and bandaging minor cuts on the job site.

Protective gear
- Wear protective clothing and foot gear (no sneakers).

- Wear face masks as appropriate.
- Wear hardhats whenever working under someone else.

Tool use and work habits
- Do not operate power equipment without ground-fault interrupters and protective devices in place
- Do not use frayed power cords.
- Use ladders carefully. Always a establish firm footing. Do not reach more than an arms' length from the ladder. Have a second person secure the ladder during ascent and descent. Do not stand on rungs marked as unsafe. Do not leave tools on top of a step ladder.

- Build scaffolding on sound footing, braced, with hand rails, mid-rails and toe boards, and with planking overlapped a minimum of 12 in.
- Keep all equipment in good condition.
- Instruct new employees on safe work habits, including proper use of equipment and proper lifting procedures.

Proper disposal methods
- Place razor blades in the disposal section of the dispenser.
- Spread out oily rags to dry so they will not ignite.

equipment breakdown, cause about 80% of job-site injuries. Often the carelessness of construction workers is a product of arrogance or of some silly macho self-image. But pride often goeth before a fall. I once worked with a builder/plumber who took great pride in his strength—he liked to display his ability to battle through and get the job done. For 20 years he succeeded in flaunting danger. Then one day he climbed to a 35-ft. high roof peak and secured himself with one powerful hand while he set a roof jack with the other. Something gave way. He skidded down the roof, over the eave, and smashed into the ground. He lived. In fact, he eventually went back to work. But he will suffer pain for the rest of his life.

Dampening arrogant disregard for safety and stirring interest in it are among the most crucial of your project-management responsibilities. Among the techniques that builders use:

- Tailgate meetings. Frequent brief meetings are likely to be more effective than occasional long ones. Workers tend to grow bored and inattentive during the long meetings and forget about safety between them. Frequent meetings keep their awareness up. At each meeting, you can discuss the main hazards associated with the work the crew will be doing in the coming days or weeks. If it's heavy form work, talk about proper lifting. If it's framing, talk about safe use of nail guns and power saws. Although you can effectively kick off a safety meeting by reading from a manual or sharing an idea from your own experience, don't lecture for the whole meeting. Instead, ask for input from the entire crew.

- Cooperative development of a company safety manual. After reading assigned portions of the OSHA safety manual, senior workers gather to pool their ideas for publication in the company policy statement.

- Safety studies and seminars. An expert in sports medicine or a similar consultant can observe your crews on site, then instruct them on safe practices. For example, a plumbing contractor's crews were experiencing frequent knee and back problems. With two days of observation and a two-hour presentation to the crew, a sports clinician virtually eliminated the problems.

- Safety inspectors. Your worker's-compensation insurer may be able to send out a site inspector who can give you tips for improving your procedures.

- Steady reminders. Whenever I visit a job site, I check it for hazards. I point out any I find to the lead, and ask that they be corrected immediately. Over and over I encourage workers not to take a chance to get something done faster; rather, I ask them to do what it takes to get it done safely.

Even beyond consciousness-raising programs, the primary source of safety at the work site is the sound management practice we have been discussing all along. When you bid fairly and accurately for clients who want good work, you have time to do the job right. Your crew has time to work safely. By contrast, if you are selling cheap work, not good, and bidding low for volume, you must cut corners. You can't afford the time for safety meetings. You must relentlessly pressure your leads to force the project forward, to meet your stingy labor estimate. Fearing for their jobs, they in turn urge their workers to hurry. In the frantic atmosphere, safety gets pushed aside—and accidents happen. The cause may show up in Department of Labor statistics as worker negligence. But often, I suspect, the negligence was forced on the workers by contractors driven to produce fast instead of well and safely.

When workers get hurt, builders pay the bill. There is, first of all, the emotional bill. How does it feel to see an employee, maybe someone who had been with you for years, dragging a crippled leg because you tried to save a nickel by skimping on scaffolding?

If the emotional bill does not deter you (and, unfortunately, there appear to be plenty of builders willing to risk employee injury for an extra buck), the financial bill should. It comes in the form of increased worker's-compensation premiums. As we have discussed, your rates are tied to your safety record. If it is good, your rate can drop by 50%. If it is bad, your rate can rise by several hundred percent. You can find it impossible to bid successfully against builders who have earned a lower rate. You can even go to jail for neglecting safety. In recent years, managers whose employees have died as a result of job-site hazards have been charged with negligent manslaughter. And they have been convicted.

JOB SETUP

Setup for a project begins with the preparation of documents. If the project requires a permit (in my territory, all but the smallest projects do), apply well in advance of the time you plan to begin construction. Make sure any preliminary paperwork required by the building department, such as insurance certificates or contractor's and business licenses, have been filed. Especially for remodeling projects, if you have not already done so during estimating, you may want to shoot a role of film of the existing conditions. Sometimes clients may unfairly attribute damage to your crew or subs, and photos can settle the dispute. Back at your office, set up the bookkeeping records you will need for the project—a page in your accounts-receivable journal, a job-cost card and unit-cost records.

A Typical Build "Do" List

As a project nears startup, Gerstel slips a Build "Do" list into his project folder. During the course of a longer project, he will update it several times.

- Make time cards
- Make flow chart
- Walk through with crew leader
- Give lead the cards and chart
- Give lead the phone number for the nearest emergency room
- Meet with subs
- Review responsibilities with owners
- Hire part-time laborers
- Check and repair all power cords
- Purchase new vibrator prior to concrete pour
- Invite Ellen from Structural Designs to see our forms; take her to lunch to explain price-planning option
- Collect change orders
- Compose the final punch list with client
- Give owner any warranties, manuals, and maintenance procedures
- Put 30/60/90 day completion calls in tickler file
- File notice of completion

For small projects, such as you might handle when you are at the beginning of your career, additional setup can probably coincide with the beginning of construction. But for larger projects, and with a larger company, you will probably want to prepare your crew leader well in advance. For my projects, I schedule a morning or even a full day to orient the lead to the project site, plans, specs, assumptions and any other documentation. We begin by walking through the site, and I describe the major items of demolition or construction and other principal features of the project. I give the lead a copy of the plans, and we discuss and highlight on them any unusual or tricky items from the foundation through the finish work. The lead will then spend additional hours studying the plans, specs and assumptions, and will take any preliminary steps—such as scheduling workers from our temporary agency—necessary for the project to get smoothly underway.

For each project, Gerstel provides the crew leader with (from left to right) plans, specifications, building notes, assumptions included in the contract and a flow chart.

During that initial walk through, I try to arrange for the client to meet the lead. Afterward, I attempt to prepare the client for construction. Communication with the client during the job should be among your major concerns. (Remember that breakdowns in communication, more than technical failures, lead to lawsuits.) When you build your clients' home, you are building their dream, a dream that is likely to put them in debt for a few decades. They are under stress and need close attention from their builder. You, in turn, need their cooperation during construction.

Often, and only half in jest, I forewarn clients of the strain of building. I remind them that they are heading into "one of the worst experiences of your life…and that's if we're lucky." I want them to brace for the noise, dirt and disruption of construction. I warn them, too, to anticipate the "90% blues." The beginning of the project, I explain, will seem to fly. After the foundation is in and the shell is up, the new house or room may seem virtually complete. Afterward, the changes will become less dramatic until, at nine-tenths of the way through the job, when small details are being completed, progress will seem to halt and the blues will set in.

Beyond urging psychological preparation, there are a number of practical comforts you can provide clients. Most important, make communication easy. Set up a "communications corner" (a fireplace mantel is always a good place), where clients can leave you information and questions, and where you can leave change orders, samples and other information. Give them the number for your lead's answering machine (presumably they already have yours), tell them the

Owner's Responsibilities Before and During Contractor's Work

As part of the assumptions included in the contract for each project, Gerstel gives his clients a list of their responsibilities (and with them an explanation of why each must be performed by a given date). Examples include:

• Have all asbestos-laden material removed before the contractor begins work

• Clear all work areas of personal possessions

• Clear the garage for tool and material storage by contractor

• Make a phone available to the contractor, crew and subs for local calls and long-distance calls, if they are job related

• Provide power and water

• Provide accessories (towel bar, toilet-tissue holder, etc., for baths and kitchen) during framing so blocking can be installed

• Provide cabinet pulls immediately after framing so cabinets can be bored

• Provide louver-door pulls after doors are installed

• Select plumbing fixtures

• Select picture molding from stock at lumber company

• The owners will complete their obligations in a timely fashion so that the contractor can proceed efficiently. The contractor will tell the owners at the beginning of the project when their tasks need to be completed. The owners will compensate the contractor for any labor or material costs that are incurred because the owners have not completed their obligations in a timely fashion. If the delay results from (or if extra work is required) as a result of errors by the supplier, we recommend that the owner back charge the supplier for our charges.

best hours to reach you, and assure them that their messages will be promptly answered. Explain to the clients that they can discuss any concerns they have with either yourself or your lead and that the two of you will keep one another posted. At the same time, remind them not to communicate directly with other workers or subcontractors, which can omit you and your lead from the communication loop, resulting in expensive confusion.

If possible, help your clients create substitutes for any facilities you will be shutting down during construction. For example, one contractor who specializes in kitchen remodeling provides a small temporary kitchen complete with storage, countertop and a microwave. With their refrigerator moved alongside, the clients can then comfortably prepare breakfast, lunches and light dinners.

The last phase of setup—creating a staging area with our tools and supplies—comes when the crew moves onto the job site. For a remodel, a portion of the clients' garage or a spare room serves our purpose. If no interior space is available, as with new construction, we can make do with tarps or rent a steel storage container.

In the staging area we install portable shelving built from 2x4s and pine planks. On one set of shelves we organize our tools, designating separate spaces for company tools and for those of each crew member. On others we place some 24 bins, many sectioned with half-gallon milk cartons, which hold a wide variety of hardware commonly needed during rough and finish work. One bin is for caulk and other compounds in tubes. Several are for joist hangers and similar framing hardware. Another contains various types of screws, tacks and staples. Still another stores different kinds of tape such as duct, plumber's and drywall. A few empty bins are reserved for hardware ordered especially for the project or that will be temporarily removed and then reinstalled. Under the shelves go 20 buckets, each holding a different type of nail. The bucket-and-bin system of storage is one of the most cost-effective of the low-overhead, low-tech investments you can make. Among its advantages are:

A number of builders who adopted Gerstel's bins-and-buckets system after reading about it have reported that it increases their efficiency.

- Gofer runs decrease. In many small-volume construction operations, gofer runs occur daily; some contractors seem to spend more time on runs to the yard than at any other single type of task. On our projects, we average about an hour a week for runs to supply yards. Our efficiency comes in part from careful advance planning and ordering of materials. But it also results from having a great variety of items at hand in our bucket-and-bin system. When an item runs low, we note it and have it shipped with our next planned delivery from the supply yard.
- Crew morale improves. Workers are frustrated when they must sift through a chaotic pile of hardware for a nail, screw or length of twine, then find they must drop their tools and run to the yard. When the item is always on hand in a well-organized storage system, they suffer no interruptions.
- Movement from project to project is eased. At a project's end, the crew can load the entire storage system into a pickup truck in an hour. At the next project, it is set up in similar time.

After job setup, before demolition or excavation begins, install any needed protection (see pp. 113-114). For new construction, protection may be minimal—a fence or tape around the work site and barriers around plants to protect them from equipment and debris. At a remodeling project, protection can be much more involved. Important items include:

- Sealing off areas of construction from lived-in areas. Six-mil polyethylene taped to the casings of doorways and to the floor does the job. If you must go back and forth through the doorway, maintain fair dust protection by installing two pieces of poly overlapped a couple of feet in the middle.
- Protecting existing floors. Double layers of red rosin paper provide basic protection. Sheets of wallboard, plywood or Masonite laid

edge to edge and taped at the seams with duct tape provide nearly complete protection.

- Protecting jambs, casings and other vulnerable finish work. Poly or red rosin paper, folded to form a pad and taped in place, is effective.
- Covering walls in passageways. Again, paper does the basic job, but sheet material is better. Sometimes protective material can be braced in place against the walls. At other times, you must tape or attach it with small fasteners and touch up when the project is complete.

With your documentation ready, your crew leader prepared, your client braced and your buckets, bins and protection in place, you are ready to build. If you have done your setup properly, rather than leaping into construction blindly, you are much more likely to have fun.

Before beginning demolition on remodeling projects, Gerstel's crews install protective coverings over all vulnerable finished surfaces.

RUNNING PROJECTS

A customer of mine who runs megaprojects for the world's largest construction company describes successful project management as "hitting all three points of the triangle." While we were remodeling his home, he liked to come in, light up a cigar and stroll around eyeballing our work. If he liked what he saw, he would say "Nice going guys, you're ringing that triangle." That meant we were doing quality work (safely), on schedule, and within the estimate.

To ring the triangle on your projects, you need good tools for organizing the myriad tasks that make up a construction job. For my projects, I have evolved a set of lists and charts to keep work in sync and to maintain clear, complete communication with clients.

The Build "Do" List has already been illustrated (p. 203). For a small job, such as construction of a porch or deck, one "Do" list might cover the entire project. For larger, more complex jobs, the list will likely need several updates and revisions; at any one time, it will cover only a fraction of all phases of the project.

For effective management, you also need to have a picture of the entire project from beginning to end. A flow chart, which lists all major phases of work in a job, gives you an overview. But the flow chart goes beyond an ordinary list, in that it tells you not only what must be done, but also when it should be done. It is the single most useful project-management tool available to small-volume builders.

The Construction-Management Triangle

Doing high-quality work on schedule and within budget is the goal of construction management.

Quality

Schedule   Estimate

To make a flow chart, you need only a sharp pencil and a large sheet of paper (I favor 11-in. by 17-in. drafting paper with grid lines). If you are at the beginning of your building career and still a little shaky on construction sequence, your first attempts to create a flow chart may be a bit of a struggle. But the effort will pay dividends and help you to understand and organize each of your jobs. As you gain experience, you will be able to create flow charts with surprising ease. Charting a complex six-month project will take an hour or less. For smaller jobs, such as a modest master bedroom/bath addition, you'll knock off the flow chart in 15 minutes—the most cost-effective minutes you will spend on the project. You can create flow charts in five steps:

1. Across the top of the chart, draw a time line showing the length of the project in weeks. (Add one or two for a grace period.) Use 0 to indicate the beginning of the job, 1 for the end of the first week, 2 for the end of the second, and so on. Because projects often don't start when planned, add actual calendar dates only when the project begins.

2. Down the left-hand side of the chart, list all major operations. Write in crew tasks in the sequence they occur. Place subcontractors in the order in which they will be needed.

3. Draw lines (shown as bars in the illustration on the facing page) to show the period during which each phase of crew or sub work will occur. The critical phases of work—namely those that must be completed before others can begin—should naturally fall along a diagonal from the upper left to the lower right corner of the chart. Noncritical items (such as finish installation by the plumber, sheet-metal sub or electrician), which can be executed at any one of several stages of the project, can be lined in outside the critical path.

4. Note that you can determine the length of the crew lines from your estimates. For example, if your crew is drawing $2,000 a week in gross wages and you have $5,000 in your estimate for framing labor, allocate 2½ weeks for framing (5,000 ÷ 2,000 = 2.5). Subcontractors should provide the amount of time they will need with their bids.

In the lower left-hand corner of the chart, block out a space for "Notifications." Here, list in proper sequence all subcontractors and suppliers who will service the project. When you contract for the project, notify each sub of his or her projected starting date, and record it on the flow chart. For example, suppose you contract for a project on February 4, 1993. You expect to begin work a month later, and to need your plumbers 1½ months into the project. You call your plumbers, tell them you expect to need them about mid April, and record the notification on the flow chart. As the job progresses and the schedule firms up, update the notifica-

A Typical Flow Chart

A flow chart is the single most valuable project-management tool. On one sheet of paper you have all the information you need to sequence operations, to schedule subtrades and to order material in advance so it does not bottleneck your project.

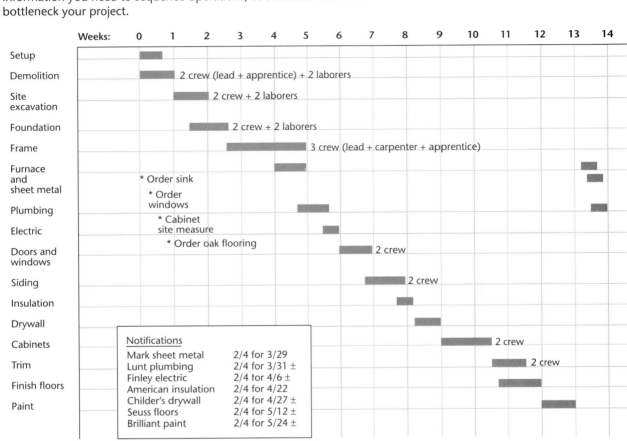

tions until you are ready to pin down each sub's starting date. Thus, at "Plumbers" you may eventually have three notifications: 2/4/93 for 3/31/93; 3/18/93 for 4/8/93; 4/6/93 for 4/14/93. With that kind of careful advance notice, the plumbers should get in right on schedule.

5. Diagonally down the chart, parallel to the time lines, note any materials that must be ordered in advance or other needs that require long lead times. For example, if the project requires custom windows, note them at the week they must be ordered. If you will need to hire an extra finish carpenter, note when to begin advertising. As a precaution, also write in any items your subs should order, so that you can give them a reminder call. A project can become seriously delayed for lack of an overlooked specialty item, such as a custom shower stall. (Since subs are involved in so many more jobs than general contractors, even the best of them can sometimes fail to procure custom items.)

With these five steps, you can produce a flow chart that gives a complete picture of your project and enables you to maintain overall control. You will be able to sequence all phases of work efficiently. Your subcontractors will be at your job site when you need them. And you will eliminate bottlenecks in your project caused by late deliveries.

With one additional step, your flow chart can become a useful tool to job-cost labor—the most volatile job cost and the one you can do most about. (With material, you are more or less stuck; you have to pay your supplier's price, so up-to-the-minute figures on costs don't enhance your management opportunities.) All you need do to track job costs with your flow chart is to write in the size and composition of the crew you estimated at each phase of work. Then, during construction, you check actual performance against the chart. That's it! As an example, in the illustration on p. 209, "3 crew (lead + carpenter + apprentice)" is written next to the time line for framing, meaning the estimate shows such a crew of three framing the project in 2½ weeks. During the project, if the planned three-person crew is framing at the expected pace, I know my labor costs are running true to the estimate. If the framing requires either more or fewer people or more or less time, I make a note on the flow chart. With the same procedure, I can track actual performance against my estimate for all phases of the work. The flow chart is so efficient a tool for keeping tabs on labor cost that, except for the largest projects, I no longer maintain a job-cost card (and thereby save myself a huge amount of boring office work).

The details of day-to-day project control require additional tools, which give a finer focus to supplement the broad overview provided by "Do" lists and flow charts. For that control, I rely on my "shirtpocket office"—the week-at-a-glance calendar and small binder notebook (shown on the facing page). An advantage of my system over larger (and fancier and costlier) organizers is its accessibility. Frequently, I have good ideas at odd moments and in odd places (like at a stop light, or during an inspection of an 18-in. high crawl space) where I could not easily lay hands on a large organizer. But shirtpocket notebooks are so handy that I can pull them out and write down my ideas before I've lost them.

I use the week-at-a-glance calendar to organize each day's work: calls to return, estimates to complete, projects to inspect, meetings to attend. I try to map out my work so that I can make the most efficient transitions between tasks and travel the shortest distance between stops. If everything won't fit in the available space, I clip in a notecard.

My small binder notebook helps me stay on top of construction details. It is divided into sections for current projects and into sub-sections for individuals—lead, client, designer— involved in each project. Each time I think of a question or comment for an individual, I write it down immediately. Then when I next talk to him or her, I open the notebook and am reminded of all the issues I need to cover. I avoid the experience of breaking through the busy signal to finally reach the building inspector, only to realize as I hang up that I forgot to mention a crucial item.

Gerstel uses two shirtpocket notebooks—a week-at-a-glance calendar and a small binder—to plan his work and keep track of details.

My leads also keep lists to help guide their day-by-day work. These lists include:

- A question sheet for each person—me, the designer, owners, subs and inspectors—involved with the project. Included in their questions for me are any possible change orders they have noted.

- Material needs. Whenever a lead or crew member secs a need for material—to execute the next phase of a project, restock a bin or handle a change order—he or she writes it on this list. The next time the lead phones in an order to a supplier or sends one of the crew on a run, all the necessary materials are obtained.

Job-Site Meetings

At their twice-weekly job-site meetings, Gerstel and each of his project leads use their documents and lists to make sure they are covering all issues. At the end of each meeting, they note all potential change orders and review and update the flow chart.

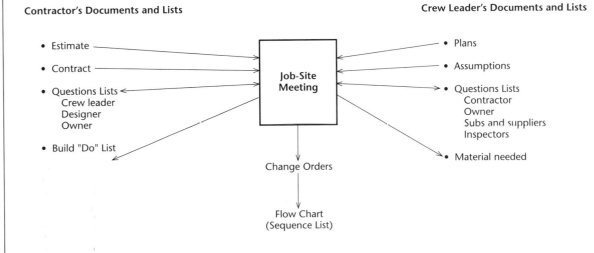

Contractor's Documents and Lists

- Estimate
- Contract
- Questions Lists
  Crew leader
  Designer
  Owner
- Build "Do" List

Job-Site Meeting

Crew Leader's Documents and Lists

- Plans
- Assumptions
- Questions Lists
  Contractor
  Owner
  Subs and suppliers
  Inspectors
- Material needed

Change Orders

Flow Chart
(Sequence List)

Crew Leader's Sequence List

When action at the job site gets too thick to track with a flow chart, the lead makes up a detailed schedule of tasks in the sequence they are to be performed:

MONDAY	TUESDAY	WEDNESDAY	THURSDAY	FRIDAY
ELECTRIC FINISH	ME- • STAIRCASE • GUARD RAIL • INSPECT ELECTRICAL WORK AND CALL WITH PUNCH LIST LESLIE - • WINDOW CASING • WINDOW STOOL • COMPLETE BASEBOARD SHEET-METAL FIN.	ME- • NEWEL CAPS • BALUSTERS LESLIE- • COVE MOLD FOR STAIR TREADS • INSTALL VANITY TOPS • CLOSET SHELVES ELECTRICIAN'S PUNCH LIST	ME- • COMPLETE BALUSTERS • TUNE-UP DOORS • WINDOW LATCHES LESLIE- • COMPLETE SHELVES • CLOSET POLES • PHONE • PATCH HOLE IN PLASTER	ME- • BUILD WINE RACK • TAKE-OFF FOR DECK • ORDER DECK MATERIAL LESLIE - • CABINET KNOBS • CAULK & SAND MITER JOINTS PLUMBING FINISH NOTIFY PAINTER

• Sequence list. Because the flow chart gives only a broad overview, it alone is sometimes inadequate to schedule complex phases of work. At such times, the lead creates a sequence list (shown above), which itemizes the tasks that must be accomplished each day by each trade. Sequence lists are especially useful for organizing the myriad details at the end of a project.

All these devices—the "Do" list and flow chart, the shirtpocket notebooks and the crew leader's lists—together form a net to catch the details of a complex project. Effective supervision of your projects, however, depends not only upon maintaining the right lists, but on integrating their use into a well-organized routine that ensures you will cover all the bases. My own job-site routine includes greeting all workers at the site, looking for safety hazards, inspecting the construction and checking it against the plans, meeting with the lead, writing change orders and meeting with the client to present change orders, discuss problems, answer questions, explain progress and collect payments.

When you arrive at a job site, you want to say hello to everyone. You are the boss, and your arrival makes the workers self-conscious. Put them at ease with a friendly greeting, and they can work more effectively. You can then swing through the site to look for hazards. Safety is of prime importance, both ethically and financially, and a separate inspection ensures that it is not pushed into the background when your attention turns to the construction itself. Be especially alert for improper use of ladders and scaffolding, or workers' laxness in the use of protective gear. Make sure that the site is clean and orderly. A well-kept site is not only safe, it's a good advertisement. Customers often cannot recognize good work, but an orderly site makes them confident that care is also being taken with construction.

After the safety tour, look over the construction and check it against the plans. Given the variety of detail in construction work, offering a checklist here is not practicable. But one item deserves special mention: flashing—at shower pans, windows, doors, decks, roof valleys and eaves, chimneys and wherever else it is needed. Improper flashing occasions nearly a third of all lawsuits against architects, and, from what I have seen, plagues small-volume builders as well. Working drawings often don't show flashing at all, and where it is shown, it may be obscurely drawn. So it falls to the builder to make sure it is done and done right. Check flashing installation repeatedly during the rough work and again at finish.

When I inspect a site, I keep my shirtpocket notebook in hand, and write down questions and ideas as they come to mind. Rather than interrupt the lead as each thought occurs, I talk to him or her after I have completed my rounds. Our discussion is organized around our lists. We start with our questions for one another. Invariably, we generate questions for other people involved in the project and think of materials that must be ordered. Each question and need is recorded on its respective list.

Next, we pull out the flow chart. We check that all necessary materials have been advance ordered and that all subcontractors have received updates of our schedule. We check whether work is moving at the expected pace with the anticipated work force, and if not, whether we can make any cost-effective adjustments: Should we fire a new apprentice who has repeatedly shown up to work tired or give him a last talking-to? Should we sub out a portion of the carpentry or schedule the subcontractors more tightly? Do we need to insist that a couple of slow subs beef up their crews and produce on their promised schedules?

If a crew is producing more slowly than I had hoped, I face one of my toughest project-management challenges: resisting the impulse to turn on the pressure. Pressure, at least in the form of anxious complaining, threatening or heavy-handed pushing, is counterproductive. Rather than provide motivation, such pressure merely creates resentment, tension and the hurrying that leads to mistakes, accidents and further losses. If you simply let the crew know that you are running behind, they will look for ways to increase production. If they cannot, you cannot ask for more. You must accept the losses, take a hard look at your estimate, and reassert to yourself and your crew that in the long run what is most important is completing the project safely and to your usual high standards of quality.

As a last step in my meeting with a project's lead, we check whether it's necessary to perform any work—any work at all—not called for in the plans or specifications and contract. If so, my next task is to write up the extras in a change order. Often, as a good-will gesture, I write up small extras at no charge, but I do make certain that every change is written up so clients remain fully informed about the project.

After meeting with the lead, I try to see the clients, or at least talk with them on the phone. I begin by reviewing the change orders. If I can't get their signatures immediately, I ask the clients to commit to signing the orders by the next day. I answer any questions and review progress, comparing actual performance against the flow chart. I tell them which subs to expect at the site in coming weeks. I go into considerable detail, explaining not only what is happening, but why. Most clients appreciate the detail and the fact that there are no surprises. That's what I'm aiming for, even with clients who grow impatient with my explanations, because surprises hurt us both.

As I talk with clients, my shirtpocket notebook reminds me of the questions I need to ask. Often I write down points they've raised that must be relayed to the lead or to others. Here the notebook does double duty as a public-relations tool—clients gain confidence when they see you immediately noting their concerns. Toward the close of our meeting, I remind clients of any payments coming due. I give them any bills for completed change orders, and collect any payments due me.

Every so often, I will encourage clients to voice any complaints they have about our performance. If there are complaints, I try to rectify matters immediately. (In occasional cases, I explain to the clients why the problem is something I must ask them to live with, at least temporarily.) With some projects, although the clients may have a few legitimate complaints, they themselves are at the heart of the problem. For no matter how carefully you qualify clients, difficult individuals occasionally slip through the screen. Maybe you find yourself working with "grinders". Or for clients who never say "Thanks" for the change orders you throw in for free, and who challenge your fair charges for extras. Some clients are irresponsible, and perpetually try to shift their obligations onto your shoulders. Still others, while diligent and fair-minded, are so nervous about the project that they will not leave you or your crew alone to build. Instead, they hover and doubt, probe and poke, and second-guess technical procedures about which they know little. They phone you constantly and at any hour. One builder I know recently received a call from an edgy client at 3:00 a.m. on a Sunday morning.

How do you handle difficult clients? I doubt there are any comprehensive answers. Your approach will depend very much on your temperament. There are, however, a few specific strategies that may help.

To begin with, you should be especially attentive to your change-order and collection procedures. Write up all change orders and have them signed before executing extra work. Collect all payments promptly and in person when due. Do not let clients run up a huge tab and get financial leverage on you. Also, politely point out to clients the clause in your contract requiring them to dispatch their responsibilities in timely fashion, and ask them to abide by it. If they do not, consider billing them for the extra labor they cost you. If the clients approach you with an unfair demand (such as putting in a drainage system not shown on the plans without charge, because "As a contractor, you should have known we would need it"), do not react immediately and risk backing them into a corner. Instead, counter with "Let me think about it" (one of the most useful phrases in the English language, at least for purposes of construction management). Often by the time you have thought about it, the client will have, too, and will have backed off. Lastly, bear in mind that although you should be patient with clients, for construction is stressful for them, you and your crew and subs are due respect and courtesy, too.

Occasionally, my crew and I have had to say "That's enough," and we have always gotten good results. For example, I arrived one day at a kitchen remodel to find that the lead—one of the most patient and forgiving people I have ever known—had ordered the client out of the room. She had been hovering over him from the first day of the project, and now she had set up a chair in a corner and sat there watching his every move, frequently leaping up to shout "No, no, no, stop." Since the lead and his helper were operating power tools, her interruptions were often dangerous as well as time-consuming and insulting. Finally, he'd had enough. He turned on the client with his big arms folded across his chest while his helper, an ex-cop who knows how to corral a suspect, closed in from the other side. "Go, and do not come back," he ordered.

When I arrived an hour later, he announced that I was going to have to fire him, told the story, then smiled when I said I felt he'd probably done the right thing. I found the client in her living room. We talked over the episode, and I explained that she was offending both the crew and me by acting as if we could not do the job properly without her supervision. She apologized, saying she realized she had pushed too far. (Like most people, she was aware of the difficult aspects of her personality. Besides, she did not want to destroy our working relationship with her kitchen half-built.) She agreed that she would inspect our work each day after we were finished, and each morning meet with the lead to discuss any problems. Otherwise, she would let us work without interruption. She did; she loves the kitchen we built for her and has referred us enthusiastically to other clients.

INSPECTORS

Inspections by the building department can provoke anxiety. You may have the feeling that these anonymous officials, who come to peer into your forms and eyeball your framing, have your fate in their hands. You may feel, too, that they can make any demand they wish, and that you will have to knuckle under. Such fears are not legitimate. Building inspectors have defined and limited power. They cannot use it capriciously. Excepting the rare incompetent, they are quite predictable, performing a routine job in a familiar pattern. In fact, though I don't suppose any of us quite get over butterflies in the stomach on inspection day, the site visit of a good inspector should be welcomed as a bulwark against potential liability problems down the road.

Building departments are hierarchies. At the top sits the building official. Beneath is the chief inspector, who supervises a staff of field inspectors and trainees, the people you see at your job sites. They come out not to appraise your work according to their own standards, but simply to enforce the building code that has been adopted by the municipality or other governing body they represent. Some municipalities write their own codes. But more typically, they take as their own one of the three model codes: the Uniform Building Code, prevalent in the western half of the U.S.; the Standard Building Code, used in the southern U.S.; and the Basic Building Code, used elsewhere.

Theoretically, the model codes are democratically produced. Anyone (including you and me) may submit proposals for changes in a code and appear at public hearings to testify on changes. In actuality, code hearings are dominated by the building officials' organization, with minor participation by trade associations, architects, engineers and manufacturers.

Building codes are not concerned with the entire spectrum of architectural and construction issues. They are much more narrowly focused, covering only matters relevant to the physical health, safety and welfare of the occupants of structures. Codes cover exits, fire protection, structural design, sanitary facilities, light and ventilation, materials, environmental issues and energy conservation.

When inspectors enforce the code, they must be concerned not only for the well being of current occupants, but for future occupants as well. Your clients may sometimes find that frustrating. "I don't want a handrail on the stairway. I don't need it, and I don't want to pay for it," they may say. But the handrail must go in. Even if the owners will never personally need it, the inspector must order the handrail installed out of concern for some future resident or guest, perhaps an elderly person or child.

It is important to realize that inspectors are not arbiters of the quality of workmanship, except as it effects health and safety issues. They will check such items as the placement of rebar, the connection of framing members and the attachment of lath—items that will determine the structural integrity of the completed building. But the quality of the finish work, which is what may matter most to clients, will lie outside their purview. Miter joints on interior window casings may open ¼ in. Drywall texture may look as if it were sprayed out of a shaving-cream can. But since inept trim and wall texture are not health and safety issues, they will garner no attention from inspectors.

Although their authority is defined and limited, inspectors do have substantial power of enforcement. Building codes have the force of any other law. If the inspector finds a code violation in your work, he or she can require you to correct it before you cover it up. If your rebar is improperly installed, you'll have to fix it before pouring concrete. If your drywall has not been screwed off at frequent enough centers, you'll have to add screws before you tape. Should you refuse to make corrections, inspectors can shut down your project. If you refuse to stop work, the inspector can call in the police and have you arrested and charged.

Challenging inspectors on their legitimate duties is not a fruitful enterprise. To work successfully with them:

- Get permits for your projects and make sure they cover the entire project. If you do not get a permit and are caught midway through the job, you may be required to tear out work so that the required inspections can be done. If you regularly try to get by without permits, or with permits that do not cover the whole scope of your project, your name will soon be mud at the building department.
- Build to the plans that have been approved by the building department. If you find during construction that you must deviate significantly from them, get the change approved before you build.
- Don't expect an inspector to be your designer or to instruct you on how to build to code. Inspectors cannot assume the liability of providing you with design and specifications.

- Know the code and build to it. Keep up with changes. Then you will not, for example, continue installing vertical railing pickets at 8 in. on center after the code has reduced the acceptable maximum to 6 in. on center

- Call for inspections when you need them—not too early and not too late. Like subcontractors, inspectors need advance notice (though the amount varies greatly among jurisdictions), and they hate being called out to a site prematurely.

- Be courteous and friendly. One building official told me courtesy made no difference, that inspectors are tough people who do not care if you are nice to them or not. Baloney. Inspectors deal with anxious, angry builders and owners all week. If they get a friendly reception at your site, they will be more inclined to work with you when your project hits one of those areas about which the code is ambiguous.

If you give inspectors what they need and do good work, you should, in turn, expect they will treat you and your crew with respect, show up on time for their appointments and enforce the code—no more and no less. They should not attempt to impose their personal ideas of sound construction. You should also expect and get careful inspections. Too often builders say of an inspector, "He's easy," as if that were a good thing. It is not. A good inspector is a thorough inspector, who makes sure the work is done to code so that down the line neither the building department, yourself nor your customers are saddled with liability problems.

You want thorough inspections, but at the same time you can reasonably expect inspectors to show some flexibility, especially on remodeling projects. The codes have been written for new construction, not for remodeling and renovation (though it now comprises the majority of construction work done in this country). Sometimes strict imposition of the code on a remodeling project creates absurdity. For example, during a kitchen remodel of mine, running the hood duct along the sensible route, straight to the nearest exterior wall, failed to achieve the full clearance required between the duct cap and an existing window by a few inches. Other routes, while meeting the letter of the code, would have required torturous runs through upstairs bedrooms. Sensibly, the inspector allowed us to go to the near outside wall and to fudge a bit on the clearance to the window.

Sometimes inspectors make a mistake. They try to impose a code that does not exist. When you cite them chapter and verse to demonstrate that your work meets the actual requirements, they should withdraw their demands. Every so often, however, you will get a problem inspector, who takes "pleasure in making people suffer," as one building official described a former colleague. He liked to stop a job by pointing out a problem, then leave, knowing that there were also other problems. When he was called to reinspect, he would point out a second problem, and on the third trip a third problem, and so on. He gloated as the project ground to a halt, awaiting his clearance. What do you do if you are confronted with a cruel or capricious inspector?

1. Ask to see the section of cited code.

2. If the citation is invalid or unreasonable because of existing conditions, point that out.

3. If the inspector refuses to yield, indicate that you will appeal to the chief inspector. Just the threat may be enough, for the inspector probably is the subject of constant complaints to the boss. In my area, there is a sheet-metal inspector notorious for abuse of power. Twice he has tried to impose unnecessarily strict requirements on our remodel projects. Both times, I've offered to relieve him of the burden of making a difficult technical decision by going to his boss. Both times, before I could get to the phone, he decided that, after all, our requests for a minor deviation from code created no real danger and that he could approve our work after all.

4. If need be, go over the head of the field inspector to the chief inspector or even the building official. Usually you will get results. Remember, you paid for your permit and for fair and thorough inspections.

5. If the building official does not respond, go further, especially when dealing with a notoriously power-mad inspector. Make a complaint—perhaps together with other builders—to the city manager and city council. Your community of responsible builders should not have to endure a bully.

 # WRAP-UP AND FOLLOW-UP

When your project has passed the final inspection, you have not necessarily reached completion. Likely, there are many finish details yet to be done. Taking care of them all, every bit of touch-up and every hardware adjustment, is critical to satisfying your clients and earning a good reference. Nothing spoils clients' enthusiasm for a job well done as quickly as a few items left dangling at the end of a project. When their friends ask, "Is your project done now?," instead of saying "Yes" with a big smile, they say "No" with a frustrated grimace, as they think of the toilet paper standing on the windowsill because the holder has not yet been installed. What does that do for your references?

The traditional tool for bringing a project to successful completion is the punch list. A punch list is exactly that, a final list of items you punch off as you do them. Traditionally, punch lists are created in one of two ways: from standard (closed) checklists, or by walking through the project and writing up an open list. Open punch lists are often made by either the designer or the client. Alternatively (and better, I think), you and your client or designer can create it together. In either case, the punch list shouldn't be drafted until you and your crew leader, working from your final "Do" lists, have dispatched as many details as possible. Your work inspires greater trust and creates a much stronger future reference when the client or designer walks through the project looking for flaws and finds hardly any.

When your clients do point out valid punch-list items, accept them without defensiveness. Do not try to explain away your oversights. And do not lay blame for them on your crew or subs. Little a manager can do is tackier. Take responsibility directly. Assure the clients, "We will take care of it." Then do so promptly.

Occasionally, at punch-list time, clients or designers make unfair requests. In particular, they will expect that whole areas of work be redone because of a small flaw. However (as your contract should specify), you do not have to replace entire cabinet doors because they have suffered a few small scratches. You do not have to repaint a room or even a wall because a few drywall nails have popped. You owe the client a careful touch-up job, and that is all.

You may want to resolve all punch-list items and have the completed list signed by the clients before they move in. Otherwise, they may attribute any damage done during the move to you or your crew or subs. If the claim seems dubious (and sometimes clients make such claims in all sincerity), you can contest it by pointing to the signed punch list.

As you close out a project, provide your clients with any manuals and manufacturers' warranties. Also give them a maintenance schedule. Let them know how often to paint those windows on the weather side to get maximum life. Remind them of the importance of cleaning gutters and changing heater filters regularly. Suggest that they start a tickler file to remind themselves when maintenance is due.

If you employ skilled workers and subs, and wrap up with carefully detailed punch lists, you should receive few callbacks. I was startled once to read a manual for builders that suggested a budget of 5% of gross receipts to cover callbacks. I can hardly imagine work done so poorly it would require that level of callback allowance. The annual cost for my company's projects has come to a small fraction of 1%.

When callbacks do come, take care of them as soon as you can. If necessary, briefly pull a lead and crew off a current job, explaining to the customers that their project will get similar treatment should the need arise. If it is truly impossible to attend to a callback right away, tell your clients exactly when you will take care of the problem, and make good on your commitment. I try to have all callbacks handled by the lead, or at least a crew member, who worked on the project. Not only is this most efficient, it is also best for client morale, because our clients know and trust their lead and crew.

Callbacks are sometimes referred to in the construction industry as the "back end of the business," a sign of the low level of consideration they often get. In fact, callbacks can be turned into an opportunity for some of the most cost-effective marketing you can do. Nothing thrills a client like your prompt attention to a problem, even though you have long ago collected your last payment. Although your legal responsibility for a project may end after a year (check the requirements in your state), if a project develops problems relatively soon after the warranty expires, you will find it worthwhile to resolve them. After all, houses and commercial structures are built to last decades, and clients should not be left holding the bag for repairs after only a couple of years. If you do not perform the repairs, especially in the event of significant failure, you may be hit with a lawsuit or complaint against your license. At the very least, you will have a bitter client, and that bitterness will spread through your reference network. Bad news moves much farther and far faster than good.

One builder I know received a call from a client three years after completing a project. A shower pan had failed. Water was seeping through the ceiling of the room below. Although the project was two years out of warranty, the builder immediately went to look at the failure and accepted responsibility for it. Soon after, he had his best tile setter on the job tearing out and rebuilding the pan. The expense of the callback proved to be inexpensive marketing. Shortly after the repair was complete, the clients referred two of their friends so emphatically, that they hired the builder without even talking to another.

RESOURCES

BOOKS

The Basic Bond Book. Washington, D.C.: Associated General Contractors of America, 1980.

Cook, Paul J. *Estimating for the General Contractor.* Kingston, Mass.: R. S. Means Co., 1982.

————. *Bidding for the General Contractor.* Kingston, Mass.: R. S. Means Co., 1985.

Guide to Construction Insurance. Washington, D.C.: Associated General Contractors of America, 1985.

Kamaroff, Bernard. *Small Time Operator.* Laytonville, Calif.: Bell Springs Publishing, 1982.

Matis, David and Jobe Toole. *Paint Contractor's Manual.* Carlsbad, Calif.: Craftsman Book Co., 1985.

Mitchell, William. *Contractor's Survival Manual.* Carlsbad, Calif.: Craftsman Book Co., 1986.

Phillips, Michael and Salli Raspberry. *Marketing Without Advertising.* Berkeley, Calif.: Nolo Press, 1986.

Schleifer, Thomas C. *Construction Contractors' Survival Guide.* New York: John Wiley & Sons, 1990.

Stoeppelwerth, Walter W. *Professional Remodeling Management.* Bethesda, Md.: Home Tech Publications, 1985.

Syvanen, Bob. *What It's Like to Build a House.* Newtown, Conn.: The Taunton Press, 1985.

Thomas, Paul I. *Estimating Tables for Homebuilding.* Carlsbad, Calif.: Craftsman Book Co., 1986.

Thomsett, Michael. *Builder's Guide to Accounting.* Carlsbad, Calif.: Craftsman Book Co., 1987.

MAGAZINES

Fine Homebuilding. The Taunton Press, 63 S. Main St., P.O. Box 5506, Newtown, Conn. 06470-5506. (800) 283-7252.

Journal of Light Construction. P.O. Box 686, Holmes, Pa. 19043. (800) 345-8112.

SAFETY MANUALS

National Safety Council. (Write for a catalog of pamphlets.) 444 N. Michigan Ave., Chicago, Ill. 60611. (312) 527-4800.

OSHA Safety and Health Standards Digest. Washington, D.C.: U.S. Department of Labor, Occupational Safety and Health Administration.

FIRST-AID SUPPLIES

Masuen, Inc., 490 Fillmore Ave., Tonawanda, N.Y. 14150. (800) 222-1934.

STORAGE EQUIPMENT

Turnkey Material Handling Co., 500 Fillmore Ave., Tonawanda, N.Y. 14150. (800) 828-7540.

ASSOCIATIONS

American Institute of Architects. 1735 New York Ave., N.W., Washington, D.C. 20006.

Associated General Contractors of America. 1957 E St., N.W., Washington, D.C. 20006.

INDEX

A

Access, to job site, 122
Accounting. *See* Bookkeeping.
Advertisements, for employees, 170-171
Advertising, 79-80
Agreements. *See* Contracts.
AIA contracts. *See* Contracts.
Answering machines, 24
Architects, working with, 93-99
Assumptions. *See* Contracts.

B

Benefits. *See* Salary and benefits.
Bid peddling, 196
Bidding:
    competitive, 99-104
    negotiated, 104-110
Bookkeepers, 71-72
Bookkeeping:
    disbursements, 53-58
    goals and tasks, 49
    overview of, 48-50
    pegboard system, 52
    pile system, 51
    receivables, 69-71
    spreadsheets, 53-55
    system for, 71-73
Breaks, lunch and coffee, 34
Budgets, of clients, 84-85
Business cards, 19, 79
Business organization, types of 15-17
Business plans. *See* Plans.

C

Callbacks, 78
Capital:
    investments, 56
    working, 35-39
Change orders, 155-157, 160-164

Clients:
    co-building with, 90-91
    dealing with, 214-215
    qualifying, 89-93
Computers:
    for estimating, 26
    office, 25
Contracts:
    agreements, 144-146
    AIA, 98, 145
    arbitration, 159-160
    assumptions, 147
    conditions, 153-155
    delay clause in, 158
    elements of, 140
    lump-sum, 140, 141
    overview of, 139-167
    payment holdback, 148-149
    payment schedule, 148
    rescission, 150-152
    rights and obligations, 157-159
    stipulations, 153-155
    subcontracts, 165-167
    terminating, 159
    time-and-materials, 140, 142-143
    types of, 140-141
Costs. *See* Estimating.
Crew:
    leaders, duties of, 204
    members, actual cost of, 177
    project orientation with, 204
    size, 176-177
    stability, 178-182
Cut list, 108

D

Design-build options, 104
Direct job costs, 53
Disbursements. *See* Bookkeeping.
Dress, for client meetings, 82

E

Estimating:
    finish-work checklist, 123
    general-conditions checklist, 118
    job cost record, 119
    labor costs, 129-131
    overall check for, 131
    overview of, 111-138
    project costs, 125-131
    rough-work checklist, 121
    site conditions, 113
    subcontractors, 130-131
    subtrade checklist, 124
    subtrades, 116
    summary sheet, 117
    unit costs, 67-69
    See also Job costing. Markup.
        Overhead.
Evaluations, of employees, 34

F

Flow chart, for projects, 208-210
Follow-up, of projects, 219-221
Formulas, for figuring material, 126-128

G

Gables, framing, 127
Graphs, for projects, 88

H

Hazards, site, 113
Hiring, 168-174. *See also* Terminating.
Holidays. *See* Salary and benefits.

I

Incorporations, 17. *See also* Business
    organization.
Inspections:
    site, 114, 212-214
    subcontractor, 114

Inspectors, 216-219
Insurance:
    additional coverage, 44
    liability, 41-44
    purchasing, 39-47
    worker's compensation, 44-47
IRA. *See* Salary and benefits.

J

Job cost card, 66
Job costing, 65-69. *See also* Estimating.

L

Labor burden, 53-54, 130.
    See also Overhead.
Liability. *See* Insurance.

M

Markup:
    fixed percentage, 135
    overview of, 132-138
    project-duration based, 136
    10 and 10, 132, 135
    See also Estimating.
Masterformat forms, 115-116.
    See also Estimating.
Medical plans, 55. *See also* Salary
    and benefits.
Meetings, job-site, 211

N

Niche, choosing, 18

O

Office:
    closet, 22
    location, 21
    portable, 27-30
    setup, 22-25
    shirt-pocket, 210-211
Organization chart, of company, 181
Overhead:
    estimating, 133-134
    fixed, 55-56
    variable, 53-54
    See also Estimating. Labor burden.

P

Partnerships, 16. *See also* Business
    organization.
Pay. *See* Salary and benefits.
Payroll, 62-65
Pegboard system. *See* Bookkeeping.
Petty cash, 58
Pile system. *See* Bookkeeping.
Plans and drawings, 86-87
Plans, business, 13-14
Policy statement, 31-35
Portfolios, photo, 83-84.
    See also Advertising.
Postcards, 81. *See also* Advertising.
Price planning, 104-110
Profit. *See* Markup
Profit sharing, 56. *See also* Salary and
    benefits.
Projects:
    graphs for, 88
    management of, 200-221
    prospective, 81-84
    qualifying, 85-89
    schedules for, 85-86
Promotion, 76-81. *See also* Advertising.
Protection, of interiors during
    construction, 206-207

Q

Quotes, phone, 130

R

Receivables. *See* Bookkeeping.
References:
    of architects, 98-99
    of employees, 173
    checking, 92
    lists of, 84
Retirement plans. *See* Salary and
    benefits.

S

Safety, 33, 200-203, 212-213
Salary and benefits:
    IRA, 188
    medical benefits, 187
    overview of,186-189
    paid holidays, 187
    profit sharing, 188-189
    retirement plans, 188
    SEP, 188
    "4/3" work schedule, 190-192
Schedules, 85-86
Setup, for jobs, 203-207
SEP. *See* Salary and benefits.
Shops, 30-31
Signs, job-site, 79
Site conditions, 113
Skills, needed by builders, 2-3
Sole proprietorships, 15. *See also*
    Business organization.
Spreadsheets. *See* Bookkeeping.
Staging areas, 205-206
Storage, of supplies, 205-206
Storage, permanent, *30*
Subcontractors and suppliers:
    qualifying, 194-196
    working with, 192-199

T

Taxes:
    filing forms, 59-61
    payroll, 64
Terminating, employees, 34,
    174-176. *See also* Hiring.
Time cards, 63
Tool allowances, 186
Tool use, 32

U

Unit costs, 68-69. *See also* Estimating.

W

Weather protection, 119
Work schedule, "4/3", 190-192
Worker's compensation. *See*
    Insurance.

Editor: Laura Tringali
Designer/layout artist: Jodie Delohery
Copy/production editor: Pam Purrone
Photographer: Metro Image Group

Typeface: ITC Stone Serif
Paper: Warren Patina matte, 70 lb., neutral pH
Printer and binder: Arcata Graphics/Hawkins, New Canton, Tennessee